Core Mathematics C3

Edexcel AS and A-level Modular Mathematics

Greg Attwood
Alistair Macpherson
Bronwen Moran
Joe Petran
Keith Pledger
Geoff Staley
Dave Wilkins

Contents

About this book

This book is designed to provide you with the best preparation possible for your Edexcel C3 unit examination:

- The LiveText CD-ROM in the back of the book contains even more resources to support you through the unit.
- A matching C3 revision guide is also available.

Brief chapter overview and 'links' to underline the importance of mathematics: to the real world, to your study of further units and to your career

Finding your way around the book

Detailed contents list shows which parts of the C3 specification are covered in each section

Every few chapters, a review exercise helps you consolidate your learning

Each section begins with a statement of what is covered in the section

Concise learning points

Step-by-step worked examples

Past examination questions are marked 'E'

Each section ends with an exercise – the questions are carefully graded so they increase in difficulty and gradually bring you up to standard

Each chapter has a different colour scheme, to help you find the right chapter quickly

Each chapter ends with a mixed exercise and a summary of key points.

At the end of the book there is an examination-style paper.

LiveText software

The LiveText software gives you additional resources: Solutionbank and Exam café. Simply turn the pages of the electronic book to the page you need, and explore!

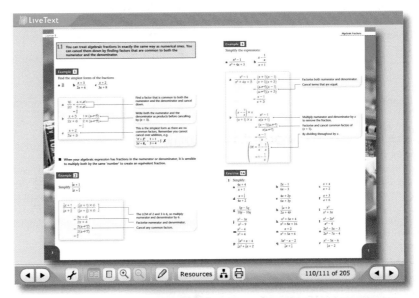

Unique Exam café feature:

- Relax and prepare – revision planner; hints and tips; common mistakes
- Refresh your memory – revision checklist; language of the examination; glossary
- Get the result! – fully worked examination-style paper

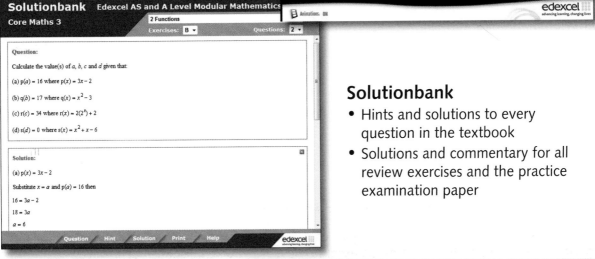

Solutionbank

- Hints and solutions to every question in the textbook
- Solutions and commentary for all review exercises and the practice examination paper

Published by Pearson Education Limited, a company incorporated in England and Wales, having its registered office at 80 Strand, London WC2R 0RL. Registered company number: 872828

www.heinemann.co.uk

Edexcel is a registered trademark of Edexcel Limited

Text © Greg Attwood, Alistair David Macpherson, Bronwen Moran, Joe Petran, Keith Pledger, Geoff Staley, David Wilkins 2008

17 16
15

British Library Cataloguing in Publication Data is available from the British Library on request.

ISBN 978 0 435519 09 4

Edited by Susan Gardner
Typeset by Tech-Set Ltd, Gateshead
Illustrated by Tech-Set Ltd, Gateshead
Cover design by Christopher Howson
Picture research by Chrissie Martin
Cover photo/illustration © Science Photo Library/Laguna Design
Printed in Italy by L.E.G.O. S.p.A

Acknowledgements
The author and publisher would like to thank the following individuals and organisations for permission to reproduce photographs:

Shutterstock/Andre Maritz p**1**; Getty Images/PhotoDisc p**12**; Shutterstock/Villiers Steyn p**31**; Shutterstock/ Michael Ledray p**45**; Photolibrary/Flirt Collection p**63**; Alamy/North Wind Picture Archives p**83**; Shutterstock/ Vitchanan Photography p**106**; Shutterstock/Bruce Amos p**132**

Every effort has been made to contact copyright holders of material reproduced in this book. Any omissions will be rectified in subsequent printings if notice is given to the publishers.

Disclaimer
This Edexcel publication offers high-quality support for the delivery of Edexcel qualifications.
Edexcel endorsement does not mean that this material is essential to achieve any Edexcel qualification, nor does it mean that this is the only suitable material available to support any Edexcel qualification. No endorsed material will be used verbatim in setting any Edexcel examination/assessment and any resource lists produced by Edexcel shall include this and other appropriate texts.

Copies of official specifications for all Edexcel qualifications may be found on the Edexcel website - www.edexcel.com

After completing this chapter you should be able to

1 'cancel down' algebraic fractions
2 multiply together two or more algebraic fractions
3 divide algebraic fractions
4 add or subtract algebraic fractions
5 convert an improper fraction into a mixed number fraction.

Algebraic fractions

Two resistors in parallel with separate resistances R_1 and R_2 have total resistance R. This can be found by using the formula

$$\frac{1}{R} = \frac{1}{R_1} + \frac{1}{R_2}$$

After completing this section you should be able to prove that the formula for R may also be expressed as

$$R = \frac{R_1 R_2}{(R_1 + R_2)}$$

The second formula is much easier to use and can only be found if you know how to manipulate algebraic fractions.

1.1 You can treat algebraic fractions in exactly the same way as numerical ones. You can cancel them down by finding factors that are common to both the numerator and the denominator.

Example **1**

Find the simplest forms of the fractions

a $\frac{16}{20}$ **b** $\frac{x+3}{2x+6}$ **c** $\frac{x+2}{3x+8}$

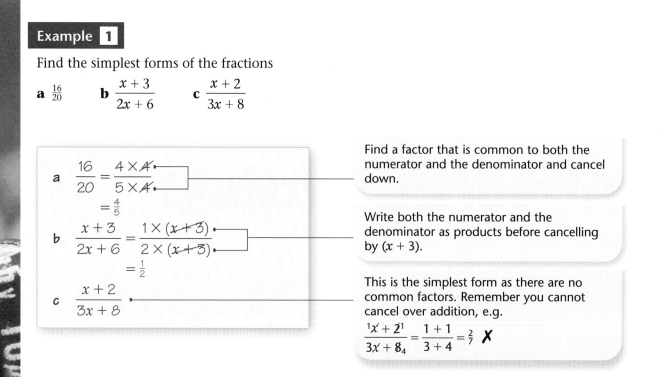

a $\dfrac{16}{20} = \dfrac{4 \times \cancel{4}}{5 \times \cancel{4}}$

$= \frac{4}{5}$

Find a factor that is common to both the numerator and the denominator and cancel down.

b $\dfrac{x+3}{2x+6} = \dfrac{1 \times \cancel{(x+3)}}{2 \times \cancel{(x+3)}}$

$= \frac{1}{2}$

Write both the numerator and the denominator as products before cancelling by $(x+3)$.

c $\dfrac{x+2}{3x+8}$

This is the simplest form as there are no common factors. Remember you cannot cancel over addition, e.g.

$\dfrac{{}^1\cancel{x} + \cancel{2}^1}{3\cancel{x} + \cancel{8}_4} = \dfrac{1+1}{3+4} = \frac{2}{7}$ ✗

■ When your algebraic expression has fractions in the numerator or denominator, it is sensible to multiply both by the same 'number' to create an equivalent fraction.

Example **2**

Simplify $\dfrac{\frac{1}{2}x + 1}{\frac{1}{3}x + \frac{2}{3}}$

$\dfrac{\frac{1}{2}x + 1}{\frac{1}{3}x + \frac{2}{3}} = \dfrac{(\frac{1}{2}x + 1) \times 6}{(\frac{1}{3}x + \frac{2}{3}) \times 6}$

$= \dfrac{3x + 6}{2x + 4}$

$= \dfrac{3\cancel{(x+2)}}{2\cancel{(x+2)}}$

$= \frac{3}{2}$

The LCM of 2 and 3 is 6, so multiply numerator and denominator by 6.

Factorise numerator and denominator.

Cancel any common factors.

Example 3

Simplify the expressions:

a $\dfrac{x^2 - 1}{x^2 + 4x + 3}$ **b** $\dfrac{x - \dfrac{1}{x}}{x + 1}$

a $\dfrac{x^2 - 1}{x^2 + 4x + 3} = \dfrac{(x + 1)(x - 1)}{(x + 1)(x + 3)}$ — Factorise both numerator and denominator.

$= \dfrac{(x + 1)(x - 1)}{(x + 1)(x + 3)}$ — Cancel terms that are equal.

$= \dfrac{x - 1}{x + 3}$

b $\dfrac{\left(x - \dfrac{1}{x}\right) \times x}{(x + 1) \times x} = \dfrac{x^2 - 1}{x(x + 1)}$ — Multiply numerator and denominator by x to remove the fraction.

$= \dfrac{(x - 1)(x + 1)}{x(x + 1)}$ — Factorise and cancel common factors of $(x + 1)$.

$= \dfrac{x - 1}{x}$ — By dividing throughout by x.

$\left(\text{OR} = \dfrac{x}{x} - \dfrac{1}{x} \right.$

$\left. = 1 - \dfrac{1}{x} \right)$

Exercise 1A

1 Simplify:

a $\dfrac{4x + 4}{x + 1}$ **b** $\dfrac{2x - 1}{6x - 3}$ **c** $\dfrac{x + 4}{x + 2}$

d $\dfrac{x + \frac{1}{2}}{4x + 2}$ **e** $\dfrac{4x + 2y}{6x + 3y}$ **f** $\dfrac{a + 3}{a + 6}$

g $\dfrac{5p - 5q}{10p - 10q}$ **h** $\dfrac{\frac{1}{2}a + b}{2a + 4b}$ **i** $\dfrac{x^2}{x^2 + 3x}$

j $\dfrac{x^2 - 3x}{x^2 - 9}$ **k** $\dfrac{x^2 + 5x + 4}{x^2 + 8x + 16}$ **l** $\dfrac{x^3 - 2x^2}{x^2 - 4}$

m $\dfrac{x^2 - 4}{x^2 + 4}$ **n** $\dfrac{x + 2}{x^2 + 5x + 6}$ **o** $\dfrac{2x^2 - 5x - 3}{2x^2 - 7x - 4}$

p $\dfrac{\frac{1}{2}x^2 + x - 4}{\frac{1}{4}x^2 + \frac{3}{2}x + 2}$ **q** $\dfrac{3x^2 - x - 2}{\frac{1}{2}x + \frac{1}{3}}$ **r** $\dfrac{x^2 - 5x - 6}{\frac{1}{3}x - 2}$

1.2 You multiply fractions together by finding the product of the numerator and dividing by the product of the denominator.

Example **4**

Calculate:

a $\frac{1}{2} \times \frac{3}{5}$ **b** $\frac{a}{b} \times \frac{c}{d}$

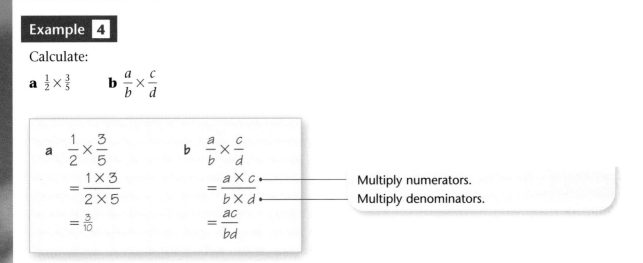

a $\frac{1}{2} \times \frac{3}{5}$

$= \frac{1 \times 3}{2 \times 5}$

$= \frac{3}{10}$

b $\frac{a}{b} \times \frac{c}{d}$

$= \frac{a \times c}{b \times d}$ Multiply numerators.
                                            Multiply denominators.

$= \frac{ac}{bd}$

■ When there are factors common to both the numerator and the denominator cancel down first.

Example **5**

Simplify the following products:

a $\frac{3}{5} \times \frac{5}{9}$ **b** $\frac{a}{b} \times \frac{c}{a}$ **c** $\frac{x+1}{2} \times \frac{3}{x^2 - 1}$

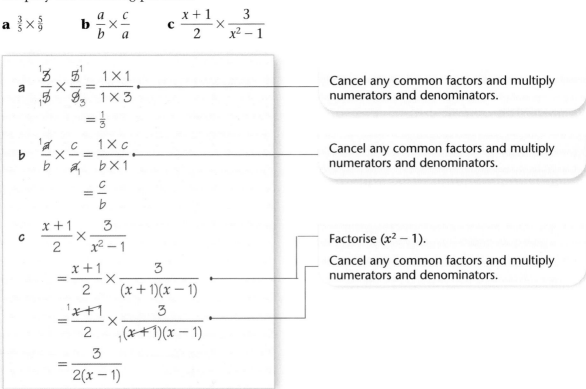

a $\dfrac{\overset{1}{\cancel{3}}}{\underset{1}{\cancel{5}}} \times \dfrac{\overset{1}{\cancel{5}}}{\underset{3}{\cancel{9}}} = \dfrac{1 \times 1}{1 \times 3}$ Cancel any common factors and multiply numerators and denominators.

$= \frac{1}{3}$

b $\dfrac{\overset{1}{\cancel{a}}}{b} \times \dfrac{c}{\underset{1}{\cancel{a}}} = \dfrac{1 \times c}{b \times 1}$ Cancel any common factors and multiply numerators and denominators.

$= \frac{c}{b}$

c $\frac{x+1}{2} \times \frac{3}{x^2 - 1}$

$= \frac{x+1}{2} \times \frac{3}{(x+1)(x-1)}$ Factorise $(x^2 - 1)$.
                                                        Cancel any common factors and multiply numerators and denominators.

$= \dfrac{\overset{1}{\cancel{x+1}}}{2} \times \dfrac{3}{\underset{1}{\cancel{(x+1)}}(x-1)}$

$= \frac{3}{2(x-1)}$

■ To divide by a fraction multiply by the reciprocal of that fraction.

Example 6

Divide $\frac{5}{6}$ by $\frac{1}{3}$.

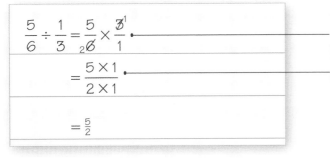

$$\frac{5}{6} \div \frac{1}{3} = \frac{5}{_2\cancel{6}} \times \frac{\cancel{3}^1}{1}$$

Turn divisor upside down and multiply by it. Cancel the common factor 3.

$$= \frac{5 \times 1}{2 \times 1}$$

Multiply numerators and denominators.

$$= \frac{5}{2}$$

Example 7

Simplify:

    a $\dfrac{a}{b} \div \dfrac{a}{c}$ **b** $\dfrac{x+2}{x+4} \div \dfrac{3x+6}{x^2-16}$

a $\dfrac{a}{b} \div \dfrac{a}{c} = \dfrac{^1\cancel{a}}{b} \times \dfrac{c}{\cancel{a}_1}$

Turn divisor upside down and multiply by it. Cancel the common factor a.

$$= \frac{1 \times c}{b \times 1}$$

Multiply numerators and denominators.

$$= \frac{c}{b}$$

b $\dfrac{x+2}{x+4} \div \dfrac{3x+6}{x^2-16}$

Turn divisor upside down and multiply by it.

$$= \frac{x+2}{x+4} \times \frac{x^2-16}{3x+6}$$

Factorise $x^2 - 16$ (difference of two squares).

Cancel any common factors.

$$= \frac{x+2}{x+4} \times \frac{(x+4)(x-4)}{3(x+2)}$$

$$= \frac{\cancel{x+2}^1}{\cancel{x+4}_1} \times \frac{(\cancel{x+4})^1(x-4)}{3(\cancel{x+2})^1}$$

$$= \frac{x-4}{3}$$

Exercise 1B

1 Simplify:

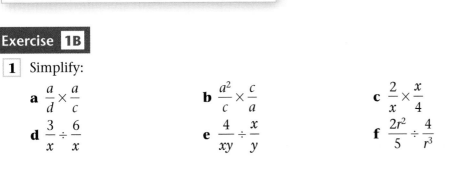

    a $\dfrac{a}{d} \times \dfrac{a}{c}$ **b** $\dfrac{a^2}{c} \times \dfrac{c}{a}$ **c** $\dfrac{2}{x} \times \dfrac{x}{4}$

    d $\dfrac{3}{x} \div \dfrac{6}{x}$ **e** $\dfrac{4}{xy} \div \dfrac{x}{y}$ **f** $\dfrac{2r^2}{5} \div \dfrac{4}{r^3}$

g $(x + 2) \times \dfrac{1}{x^2 - 4}$

h $\dfrac{1}{a^2 + 6a + 9} \times \dfrac{a^2 - 9}{2}$

i $\dfrac{x^2 - 3x}{y^2 + y} \times \dfrac{y + 1}{x}$

j $\dfrac{y}{y + 3} \div \dfrac{y^2}{y^2 + 4y + 3}$

k $\dfrac{x^2}{3} \div \dfrac{2x^3 - 6x^2}{x^2 - 3x}$

l $\dfrac{4x^2 - 25}{4x - 10} \div \dfrac{2x + 5}{8}$

m $\dfrac{x + 3}{x^2 + 10x + 25} \times \dfrac{x^2 + 5x}{x^2 + 3x}$

n $\dfrac{3y^2 + 4y - 4}{10} \div \dfrac{3y + 6}{15}$

o $\dfrac{x^2 + 2xy + y^2}{2} \times \dfrac{4}{(x - y)^2}$

1.3 **You need to have the same denominator when you add or subtract numeric or algebraic fractions.**

Example **8**

Add $\frac{1}{3}$ to $\frac{3}{4}$.

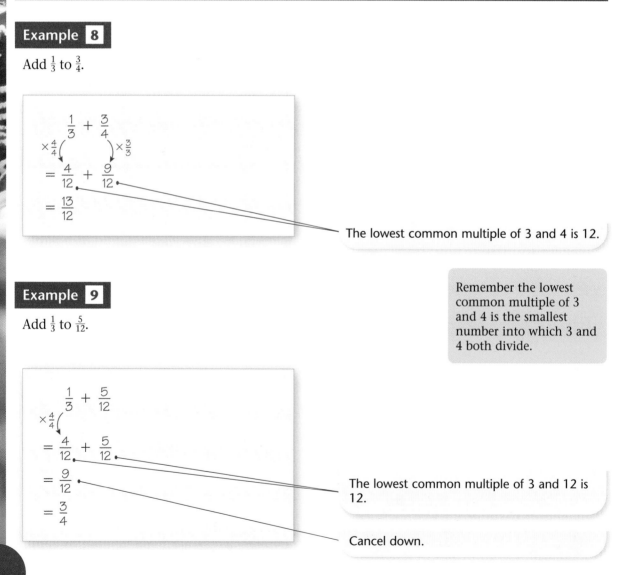

The lowest common multiple of 3 and 4 is 12.

Example **9**

Add $\frac{1}{3}$ to $\frac{5}{12}$.

Remember the lowest common multiple of 3 and 4 is the smallest number into which 3 and 4 both divide.

The lowest common multiple of 3 and 12 is 12.

Cancel down.

Example 10

Simplify the following fractions:

a $\dfrac{a}{x} + b$

b $\dfrac{3}{x + 1} - \dfrac{4x}{x^2 - 1}$

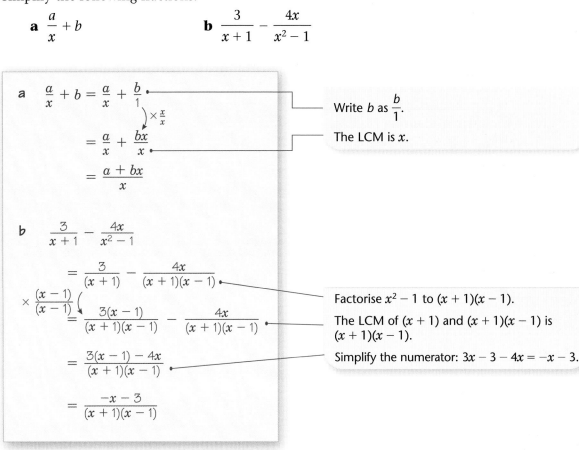

a $\dfrac{a}{x} + b = \dfrac{a}{x} + \dfrac{b}{1}$

$\qquad\qquad\times \frac{x}{x}$

Write b as $\dfrac{b}{1}$.

$\qquad = \dfrac{a}{x} + \dfrac{bx}{x}$

The LCM is x.

$\qquad = \dfrac{a + bx}{x}$

b $\dfrac{3}{x + 1} - \dfrac{4x}{x^2 - 1}$

$\qquad = \dfrac{3}{(x + 1)} - \dfrac{4x}{(x + 1)(x - 1)}$

$\times \dfrac{(x - 1)}{(x - 1)}$

$\qquad = \dfrac{3(x - 1)}{(x + 1)(x - 1)} - \dfrac{4x}{(x + 1)(x - 1)}$

Factorise $x^2 - 1$ to $(x + 1)(x - 1)$.

The LCM of $(x + 1)$ and $(x + 1)(x - 1)$ is $(x + 1)(x - 1)$.

$\qquad = \dfrac{3(x - 1) - 4x}{(x + 1)(x - 1)}$

Simplify the numerator: $3x - 3 - 4x = -x - 3$.

$\qquad = \dfrac{-x - 3}{(x + 1)(x - 1)}$

Exercise 1C

1 Simplify:

a $\dfrac{1}{p} + \dfrac{1}{q}$

b $\dfrac{a}{b} - 1$

c $\dfrac{1}{2x} + \dfrac{1}{x}$

d $\dfrac{3}{x^2} - \dfrac{1}{x}$

e $\dfrac{3}{4x} + \dfrac{1}{8x}$

f $\dfrac{x}{y} + \dfrac{y}{x}$

g $\dfrac{1}{x + 2} - \dfrac{1}{x + 1}$

h $\dfrac{2}{x + 3} - \dfrac{1}{x - 2}$

i $\frac{1}{3}(x + 2) - \frac{1}{2}(x + 3)$

j $\dfrac{3x}{(x + 4)^2} - \dfrac{1}{(x + 4)}$

k $\dfrac{1}{2(x + 3)} + \dfrac{1}{3(x - 1)}$

l $\dfrac{2}{x^2 + 2x + 1} + \dfrac{1}{x + 1}$

m $\dfrac{3}{x^2 + 3x + 2} - \dfrac{2}{x^2 + 4x + 4}$

n $\dfrac{2}{a^2 + 6a + 9} - \dfrac{3}{a^2 + 4a + 3}$

o $\dfrac{2}{y^2 - x^2} + \dfrac{3}{y - x}$

p $\dfrac{x + 2}{x^2 - x - 12} - \dfrac{x + 1}{x^2 + 5x + 6}$

q $\dfrac{3x + 1}{(x + 2)^3} - \dfrac{2}{(x + 2)^2} + \dfrac{4}{(x + 2)}$

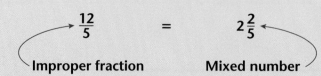

1.4 You can divide two algebraic fractions. As with ordinary numbers, an improper fraction has a larger numerator than denominator, e.g.

$$\frac{12}{5} \qquad = \qquad 2\frac{2}{5}$$

Improper fraction   Mixed number

An improper algebraic fraction is one whose numerator has a degree equal to or larger than the denominator, e.g.

$$\frac{x^2 + 5x + 8}{x - 2} \quad \text{and} \quad \frac{x^3 - 8x + 5}{x^2 + 2x + 1}$$

are both improper fractions because the numerator has a larger degree than the denominator. You can change them into mixed number fractions either by long division or by using the remainder theorem.

Example 11

Divide $x^3 + x^2 - 7$ by $x - 3$ by using long division.

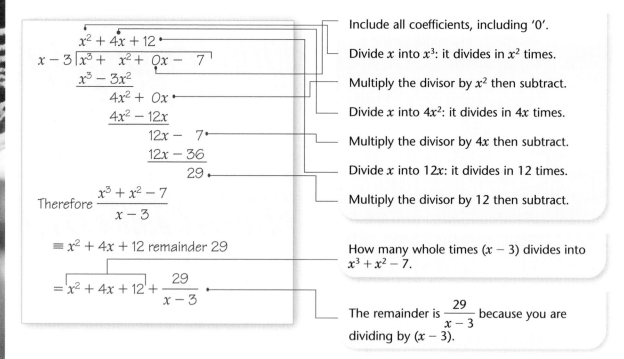

Include all coefficients, including '0'.

Divide x into x^3: it divides in x^2 times.

Multiply the divisor by x^2 then subtract.

Divide x into $4x^2$: it divides in $4x$ times.

Multiply the divisor by $4x$ then subtract.

Divide x into $12x$: it divides in 12 times.

Multiply the divisor by 12 then subtract.

How many whole times $(x - 3)$ divides into $x^3 + x^2 - 7$.

The remainder is $\dfrac{29}{x - 3}$ because you are dividing by $(x - 3)$.

■ **The remainder theorem:**
Any polynomial F(x) can be put in the form

$$\text{F}(x) \equiv \text{Q}(x) \times \text{divisor} + \text{remainder}$$

where Q(x) is the quotient and is how many times the divisor divides into the function.

This is a more general version of the remainder theorem you met in Book C2. It allows you to divide by a quadratic expression.

Example 12

Divide $x^3 + x^2 - 7$ by $x - 3$ by using the remainder theorem.

As the divisor is a linear expression and F(x) is a cubic polynomial then Q(x) must be a quadratic and the remainder must be a constant.

Set up the identity

$F(x) \equiv Q(x) \times \text{divisor} + \text{remainder}$
$x^3 + x^2 - 7 \equiv (Ax^2 + Bx + C)(x - 3) + D$

and solve to find the constants A, B, C and D.

This is true for all x, so you can substitute into the RHS and LHS to work out the values of A, B, C and D.

Let $x = 3$

$$27 + 9 - 7 = (9A + 3B + C) \times 0 + D$$
$$D = 29$$

Let $x = 0$

$$0 + 0 - 7 = (A \times 0 + B \times 0 + C) \times (0 - 3) + D$$
$$-7 = -3C + D$$
$$-7 = -3C + 29$$
$$-3C = -36$$
$$C = 12$$

Put $x = 3$ into both sides to give an equation in D only.

Substitute $D = 29$ and $x = 0$ to give an equation in C only.

Compare coefficients in x^3 $\quad 1 = A$

Compare coefficients in x^2 $\quad 1 = -3A + B$
$$1 = -3 + B$$
$$B = 4$$

Because this is an equivalence relation you can compare coefficients of terms in x^3 and x^2 on the RHS and LHS of the equation

Therefore

$$x^3 + x^2 - 7$$
$$\equiv (1x^2 + 4x + 12)(x - 3) + 29$$

$$\Rightarrow \frac{x^3 + x^2 - 7}{x - 3} \equiv x^2 + 4x + 12 + \frac{29}{x - 3}$$

In x^3: LHS $= x^3$
$\qquad$ RHS $= Ax^3$
In x^2: LHS $= x^2$
$\qquad$ RHS $= (-3A + B)x^2$

Substitute $A = 1$.

$\div (x - 3)$

Example 13

Divide $x^4 + x^3 + x - 10$ by $x^2 + 2x - 3$.

Method 1. Using long division.

$$
\begin{array}{r}
x^2 - x + 5 \\
x^2 + 2x - 3 \overline{\smash{)}x^4 + x^3 + 0x^2 + x - 10} \\
\underline{x^4 + 2x^3 - 3x^2} \\
-x^3 + 3x^2 + x \\
\underline{-x^3 - 2x^2 + 3x} \\
5x^2 - 2x - 10 \\
\underline{5x^2 + 10x - 15} \\
-12x + 5
\end{array}
$$

All coefficients need to be included.

x^2 goes into x^4, x^2 times. Multiply x^2 by $(x^2 + 2x - 3)$ and subtract.

x^2 goes into $-x^3$, $-x$ times. Multiply $-x$ by $(x^2 + 2x - 3)$ and subtract.

x^2 goes into $5x^2$, 5 times. Multiply 5 by $(x^2 + 2x - 3)$ and subtract.

Since the degree of $(-12x + 5)$ is smaller than $(x^2 + 2x - 3)$ this is the remainder.

Therefore

$$\frac{x^4 + x^3 + x - 10}{x^2 + 2x - 3}$$

$$= x^2 - x + 5 \text{ remainder } -12x + 5$$

$$= x^2 - x + 5 + \frac{-12x + 5}{x^2 + 2x - 3}$$ •————— Write the remainder as a 'fraction'.

Method 2. Using the remainder theorem.

$F(x) \equiv Q(x) \times \text{divisor} + \text{remainder}$
$x^4 + x^3 + x - 10$
$\equiv (Ax^2 + Bx + C)(x^2 + 2x - 3) + Dx + E$ •———

Now solve for A, B, C, D and E. •——————

As the divisor is a quadratic expression and $F(x)$ has a power of 4 then $Q(x)$ must be a quadratic and the remainder must be a linear expression.

As $x^2 + 2x - 3 \equiv (x + 3)(x - 1)$ we should start by substituting $x = 1$, $x = -3$ then $x = 0$.

Exercise **1D**

1 Express the following improper fractions in 'mixed' number form by:
   i using long division   **ii** using the remainder theorem

a $\dfrac{x^3 + 2x^2 + 3x - 4}{x - 1}$

b $\dfrac{2x^3 + 3x^2 - 4x + 5}{x + 3}$

c $\dfrac{x^3 - 8}{x - 2}$

d $\dfrac{2x^2 + 4x + 5}{x^2 - 1}$

e $\dfrac{8x^3 + 2x^2 + 5}{2x^2 + 2}$

f $\dfrac{4x^3 - 5x^2 + 3x - 14}{x^2 + 2x - 1}$

g $\dfrac{x^4 + 3x^2 - 4}{x^2 + 1}$

h $\dfrac{x^4 - 1}{x + 1}$

i $\dfrac{2x^4 + 3x^3 - 2x^2 + 4x - 6}{x^2 + x - 2}$

2 Find the value of the constants A, B, C, D and E in the following identity:

$$3x^4 - 4x^3 - 8x^2 + 16x - 2 \equiv (Ax^2 + Bx + C)(x^2 - 3) + Dx + E$$

Mixed exercise **1E**

1 Simplify the following fractions:

a $\dfrac{ab}{c} \times \dfrac{c^2}{a^2}$

b $\dfrac{x^2 + 2x + 1}{4x + 4}$

c $\dfrac{x^2 + x}{2} \div \dfrac{x + 1}{4}$

d $\dfrac{x + \dfrac{1}{x} - 2}{x - 1}$

e $\dfrac{a + 4}{a + 8}$

f $\dfrac{b^2 + 4b - 5}{b^2 + 2b - 3}$

2 Simplify:

a $\dfrac{x}{4} + \dfrac{x}{3}$

b $\dfrac{4}{y} - \dfrac{3}{2y}$

c $\dfrac{x + 1}{2} - \dfrac{x - 2}{3}$

d $\dfrac{x^2 - 5x - 6}{x - 1}$

e $\dfrac{x^3 + 7x - 1}{x + 2}$

f $\dfrac{x^4 + 3}{x^2 + 1}$

3 Find the value of the constants A, B, C and D in the following identity:

$$x^3 - 6x^2 + 11x + 2 \equiv (x - 2)(Ax^2 + Bx + C) + D$$

4 $f(x) = x + \dfrac{3}{x - 1} - \dfrac{12}{x^2 + 2x - 3}$ $\{x \in \mathbb{R}, x > 1\}$

Show that $f(x) = \dfrac{x^2 + 3x + 3}{x + 3}$

E

5 Show that $\dfrac{x^4 + 2}{x^2 - 1} \equiv x^2 + B + \dfrac{C}{x^2 - 1}$ for constants B and C, which should be found.

6 Show that $\dfrac{4x^3 - 6x^2 + 8x - 5}{2x + 1}$ can be put in the form $Ax^2 + Bx + C + \dfrac{D}{2x + 1}$. Find the

values of the constants A, B, C and D.

Summary of key points

1 Algebraic fractions can be simplified by cancelling down. To do this the numerators and denominators must be fully factorised first.

2 If the numerator and denominator contain fractions then you can multiply both by the same number (the lowest common multiple) to create an equivalent fraction.

3 To multiply fractions, you simply multiply the numerators and multiply the denominators. If possible cancel down first.

4 To divide two fractions, multiply the first fraction by the reciprocal of the second fraction.

5 To add (or subtract) fractions each fraction must have the same denominator. This is done by finding the **lowest** common multiple of the denominators.

6 When the numerator has the same or higher degree than the denominator, you can divide the terms to produce a 'mixed' number fraction. This can be done either by using long division or by using the remainder theorem:

$$F(x) \equiv Q(x) \times \text{divisor} + \text{remainder}$$

where $Q(x)$ is the quotient and is how many times the divisor divides into the function.

2

Functions

After completing this chapter you should be able to

1 represent a mapping by a diagram, by an equation and by a graph

2 understand the terms function, domain and range

3 combine two or more functions to make a composite function

4 know the difference between a 'one to one' and 'many to one' function

5 know how to find the inverse of a function

6 know the relationship between the graph of a function and its inverse.

There are many examples of functions in real life. One interesting case involves electricity charges where you are charged different amounts per unit dependent upon how many units you use. This is a typical case:

'The charge is 10p per kW hour up to and including a usage of 250 kW hours and 8p per kW hour after that'.

The electricity bill for this house would be high!

2.1 A mapping transforms one set of numbers into a different set of numbers. The mapping can be described in words or through an algebraic equation. It can also be represented by a Cartesian graph.

Example 1

Draw mapping diagrams and graphs for the following operations:

a 'add 3' on the set $\{-3, 1, 4, 6, x\}$ **b** 'square' on the set $\{-1, 1, -2, 2, x\}$

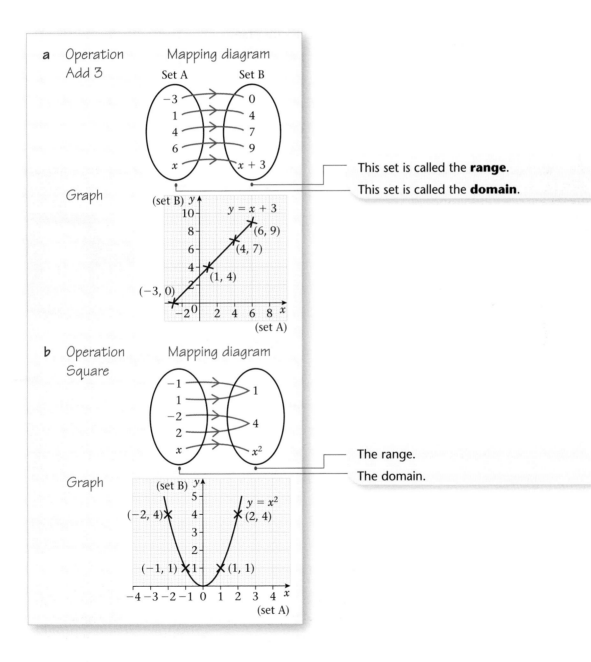

a Operation Add 3

Mapping diagram

Set A → Set B

This set is called the **range**.

This set is called the **domain**.

Graph

(set B) $y = x + 3$

(6, 9)
(4, 7)
(1, 4)
(−3, 0)

(set A)

b Operation Square

Mapping diagram

The range.

The domain.

Graph

(set B) $y = x^2$

(−2, 4) (2, 4)
(−1, 1) (1, 1)

(set A)

Exercise 2A

1 Draw mapping diagrams and graphs for the following operations:

   a 'subtract 5' on the set $\{10, 5, 0, -5, x\}$

   b 'double and add 3' on the set $\{-2, 2, 4, 6, x\}$

   c 'square and then subtract 1' on the set $\{-3, -1, 0, 1, 3, x\}$

   d 'the positive square root' on the set $\{-4, 0, 1, 4, 9, x\}$.

2 Find the missing numbers **a** to **h** in the following mapping diagrams:

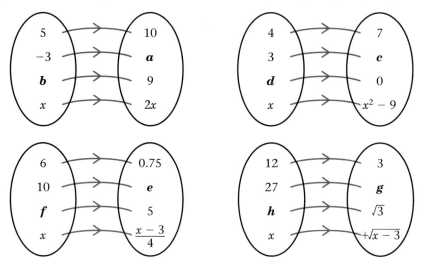

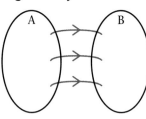 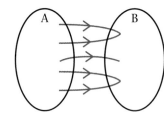 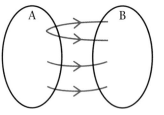

2.2 A function is a special mapping such that every element of set A (the domain) is mapped to exactly one element of set B (the range).

■ A good way to remember this is

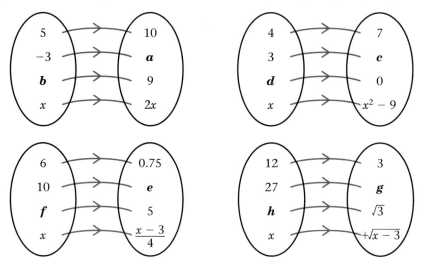

      one-to-one   many-to-one   not a function
       function   function

You can write functions in two different ways:

    $f(x) = 2x + 1$ OR $f : x \rightarrow 2x + 1$

> This is the function 'double and add 1'.

Example 2

Given that the function $g(x) = 2x^2 + 3$, find:

a the value of $g(2)$

b the value of a such that $g(a) = 35$

c the range of the function.

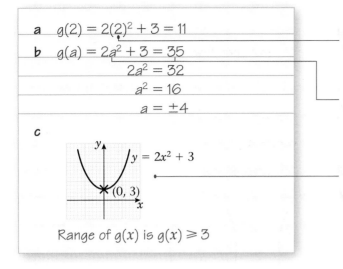

a $g(2) = 2(2)^2 + 3 = 11$ — Substitute $x = 2$ in the formula.

b $g(a) = 2a^2 + 3 = 35$
$$2a^2 = 32$$
$$a^2 = 16$$
$$a = \pm 4$$
— Substitute $x = a$ and $g(a) = 35$.

c
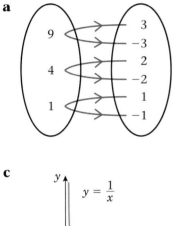

$y = 2x^2 + 3$

$(0, 3)$

Sketch the function $g(x)$. The range is the values that y takes.

Range of $g(x)$ is $g(x) \geqslant 3$

Example 3

Which of the following mappings represent functions? Give reasons for your answers.

a
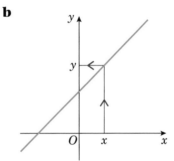

9

4

1

3
−3
2
−2
1
−1

b

c
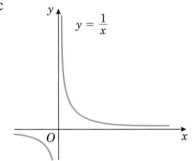

$y = \dfrac{1}{x}$

d
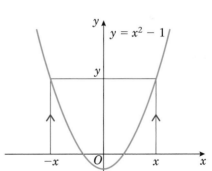

$y = x^2 - 1$

a Every element of set A gets mapped to two elements in set B. Therefore this is _not_ a function. It is an operation (square root).

b Every value of x gets mapped to one value of y. This is therefore a function. It can be written in the form $y = ax + b$ for some constants a and b. This type of function is called

one-to-one because every element of the range comes exactly from one element in the domain.

c   On the sketch of $y = \frac{1}{x}$ you can see that $x = 0$ does not get mapped anywhere. Therefore not all elements of set A get mapped to elements in set B. Hence it is *not* a function.

d   This is a function. It is called a **many-to-one** function because two elements of the domain get mapped to one element in the range.

Exercise 2B

1   Find:

a   f(3) where $f(x) = 5x + 1$
b   g(−2) where $g(x) = 3x^2 − 2$
c   h(0) where $h : x \rightarrow 3^x$
d   j(−2) where $j : x \rightarrow 2^{-x}$

2   Calculate the value(s) of a, b, c and d given that:

a   p(a) = 16 where $p(x) = 3x − 2$
b   q(b) = 17 where $q(x) = x^2 − 3$
c   r(c) = 34 where $r(x) = 2(2^x) + 2$
d   s(d) = 0 where $s(x) = x^2 + x − 6$

3   For the following functions

i   sketch the graph of the function
ii   state the range
iii   describe if the function is one-to-one or many-to-one.

a $m(x) = 3x + 2$ b $n(x) = x^2 + 5$ c $p(x) = \sin(x)$ d $q(x) = x^3$

4   State whether or not the following graphs represent functions. Give reasons for your answers and describe the type of function.

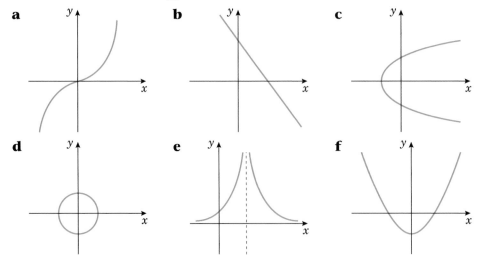

2.3 Many mappings can be made into functions by changing the domain.

Consider $y = \sqrt{x}$

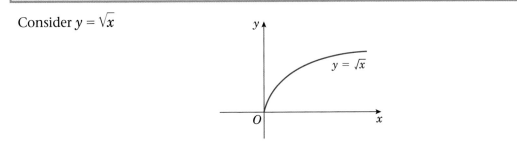

If the domain is all of the real numbers $\{x \in \mathbb{R}\}$, then this is not a function because values of x less than 0 do not get mapped anywhere.

If you restrict the domain to $x \geqslant 0$, all of set A gets mapped to exactly one element of set B.

We can write this function $f(x) = \sqrt{x}$, with domain $\{x \in \mathbb{R}, x \geqslant 0\}$.

Example 4

Find the range of the following functions:

a $f(x) = 3x - 2$, domain $\{x = 1, 2, 3, 4\}$ **b** $g(x) = x^2$, domain $\{x \in \mathbb{R}, -5 \leqslant x \leqslant 5\}$

c $h(x) = \dfrac{1}{x}$ domain $\{x \in \mathbb{R}, 0 < x \leqslant 3\}$

State if the functions are one-to-one or many-to-one.

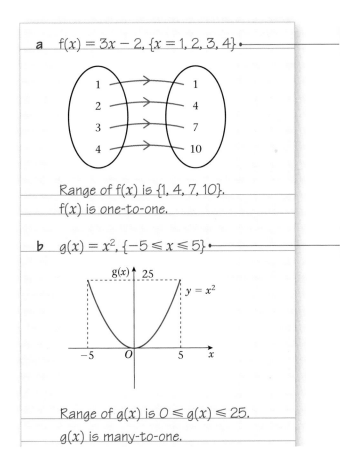

a $f(x) = 3x - 2$, $\{x = 1, 2, 3, 4\}$

Here the domain is discrete as it only has integer values. Draw a mapping diagram.

Range of $f(x)$ is $\{1, 4, 7, 10\}$.

$f(x)$ is one-to-one.

b $g(x) = x^2$, $\{-5 \leqslant x \leqslant 5\}$

Here the domain is continuous. It takes all values between -5 and 5. Sketch a graph.

Range of $g(x)$ is $0 \leqslant g(x) \leqslant 25$.

$g(x)$ is many-to-one.

c $h(x) = \dfrac{1}{x}$ $\{x \in \mathbb{R}, 0 < x \leqslant 3\}$

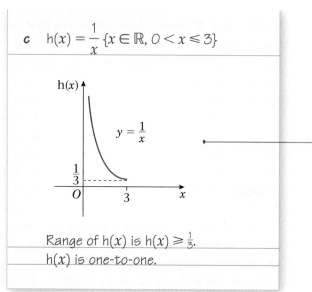

$y = \dfrac{1}{x}$

Sketch a graph.

Range of $h(x)$ is $h(x) \geqslant \frac{1}{3}$.

$h(x)$ is one-to-one.

Example 5

The function $f(x)$ is defined by

$$f(x) = \begin{cases} 5 - 2x & x < 1 \\ x^2 + 3 & x \geqslant 1 \end{cases}$$

a Sketch $f(x)$ stating its range.

b Find the values of a such that $f(a) = 19$.

Note: This function consists of two parts – one linear (for $x < 1$), the other quadratic (for $x \geqslant 1$). A useful tip in drawing the function is to sketch both parts separately and also to find the value of both parts at $x = 1$.

a $f(1)$, using $f(x) = 5 - 2x = 5 - 2 = 3$

$f(1)$, using $f(x) = x^2 + 3 = 1 + 3 = 4$

For $x < 1$, $f(x)$ is linear. It has gradient -2 and passes through 5 on the y axis.

For $x \geqslant 1$, $f(x)$ is a $\bigvee$-shaped quadratic. You need to calculate the value of $f(1)$ on both curves.

The range is the values that y takes and therefore $f(x) > 3$.

Note that $f(x) \neq 3$ at $x = 1$

so $f(x) > 3$

not $f(x) \geqslant 3$

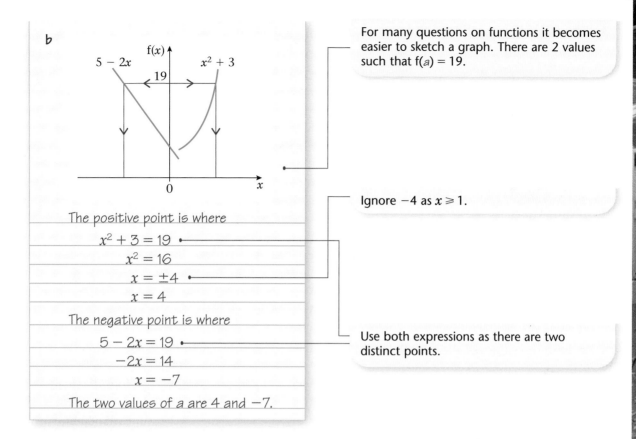

b

The positive point is where

$$x^2 + 3 = 19$$
$$x^2 = 16$$
$$x = \pm 4$$
$$x = 4$$

The negative point is where

$$5 - 2x = 19$$
$$-2x = 14$$
$$x = -7$$

The two values of a are 4 and -7.

For many questions on functions it becomes easier to sketch a graph. There are 2 values such that $f(a) = 19$.

Ignore -4 as $x \geqslant 1$.

Use both expressions as there are two distinct points.

Exercise 2C

1 The functions below are defined for the discrete domains.

 i Represent each function on a mapping diagram, writing down the elements in the range.

 ii State if the function is one-to-one or many-to-one.

   a $f(x) = 2x + 1$ for the domain $\{x = 1, 2, 3, 4, 5\}$.

   b $g(x) = +\sqrt{x}$ for the domain $\{x = 1, 4, 9, 16, 25, 36\}$.

   c $h(x) = x^2$ for the domain $\{x = -2, -1, 0, 1, 2\}$.

   d $j(x) = \dfrac{2}{x}$ for the domain $\{x = 1, 2, 3, 4, 5\}$.

2 The functions below are defined for continuous domains.

 i Represent each function on a graph.

 ii State the range of the function.

 iii State if the function is one-to-one or many-to-one.

   a $m(x) = 3x + 2$ for the domain $\{x > 0\}$.

   b $n(x) = x^2 + 5$ for the domain $\{x \geqslant 2\}$.

   c $p(x) = 2\sin x$ for the domain $\{0 \leqslant x \leqslant 180\}$.

   d $q(x) = +\sqrt{x + 2}$ for the domain $\{x \geqslant -2\}$.

3 The following mappings f and g are defined on all the real numbers by

$$f(x) = \begin{cases} 4 - x & x < 4 \\ x^2 + 9 & x \geq 4 \end{cases} \qquad g(x) = \begin{cases} 4 - x & x < 4 \\ x^2 + 9 & x > 4 \end{cases}$$

Explain why $f(x)$ is a function and $g(x)$ is not.
Sketch the function $f(x)$ and find

a $f(3)$

b $f(10)$

c the value(s) of a such that $f(a) = 90$.

4 The function s is defined by

$$s(x) = \begin{cases} x^2 - 6 & x < 0 \\ 10 - x & x \geq 0 \end{cases}$$

a Sketch $s(x)$.

b Find the value(s) of a such that $s(a) = 43$.

c Find the values of the domain that get mapped to themselves in the range.

5 The function g is defined by $g(x) = cx + d$ where c and d are constants to be found. Given $g(3) = 10$ and $g(8) = 12$ find the values of c and d.

6 The function f is defined by $f(x) = ax^3 + bx - 5$ where a and b are constants to be found. Given that $f(1) = -4$ and $f(2) = 9$, find the values of the constants a and b.

7 The function h is defined by $h(x) = x^2 - 6x + 20 \ \{x \geq a\}$.
Given that $h(x)$ is a one-to-one function find the smallest possible value of the constant a.

> **Hint:** Complete the square for $h(x)$.

2.4 You can combine two or more basic functions to make a new more complex function.

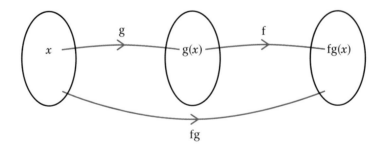

■ **fg(x)** means apply g first, followed by f. fg(x) is called a **composite function**.

Example **6**

Given $f(x) = x^2$ and $g(x) = x + 1$, find:

a fg(1)   **b** fg(3)   **c** fg(x)

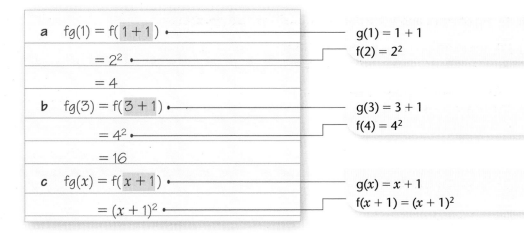

a $fg(1) = f(\boxed{1 + 1})$ ———————— $g(1) = 1 + 1$

  $= 2^2$ ———————————— $f(2) = 2^2$

  $= 4$

b $fg(3) = f(\boxed{3 + 1})$ ———————— $g(3) = 3 + 1$

  $= 4^2$ ———————————— $f(4) = 4^2$

  $= 16$

c $fg(x) = f(\boxed{x + 1})$ ———————— $g(x) = x + 1$

  $= (x + 1)^2$ ———————— $f(x + 1) = (x + 1)^2$

Example 7

The functions f and g are defined by $f(x) = 3x + 2$ and $g(x) = x^2 + 4$. Find:

a the function $fg(x)$

b the function $gf(x)$

c the function $f^2(x)$

Note: $f^2(x)$ is $ff(x)$.

d the values of b such that $fg(b) = 62$

a $fg(x) = f(\boxed{x^2 + 4})$ ———————— g acts on x first, mapping it to $x^2 + 4$.

  $= 3(x^2 + 4) + 2$ ———————— f acts on the result. f 'trebles and then adds 2'.

  $= 3x^2 + 14$ ———————— Simplify answer.

b $gf(x) = g(\boxed{3x + 2})$ ———————— f acts on x first, mapping it to $3x + 2$.

  $= (3x + 2)^2 + 4$ ————————— g acts on the result. g 'squares and then adds 4'.

  $= 9x^2 + 12x + 8$

  Simplify answer.

c $f^2(x) = f(\boxed{3x + 2})$ ———————— f maps x to $3x + 2$.

  $= 3(3x + 2) + 2$ ————————— f acts on the result. f 'trebles numbers then adds 2'.

  $= 9x + 8$

d $fg(x) = 3x^2 + 14$

  If $fg(b) = 62$

  $3b^2 + 14 = 62$ ———————— Set up and solve an equation in b.

  $3b^2 = 48$

  $b^2 = 16$

  $b = \pm 4$

Example 8

The functions m(x), n(x) and p(x) are defined by $m(x) = \dfrac{1}{x}$, $n(x) = 2x + 4$, $p(x) = x^2 - 2$.
Find in terms of m, n and p the functions

a $\dfrac{2}{x} + 4$ **b** $4x^2 + 16x + 14$ **c** $\dfrac{1}{4x + 12}$

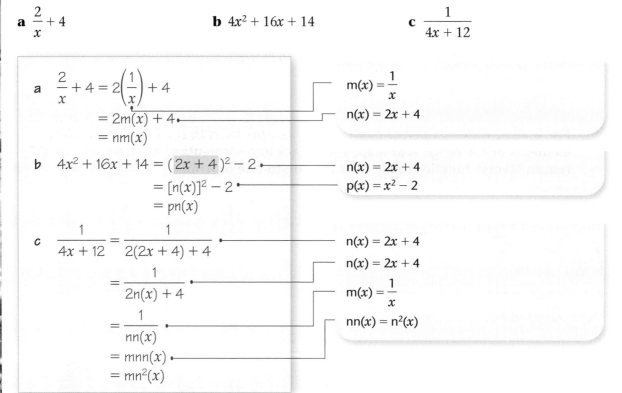

a $\dfrac{2}{x} + 4 = 2\left(\dfrac{1}{x}\right) + 4$ $m(x) = \dfrac{1}{x}$

$= 2m(x) + 4$ $n(x) = 2x + 4$

$= nm(x)$

b $4x^2 + 16x + 14 = (2x + 4)^2 - 2$ $n(x) = 2x + 4$

$= [n(x)]^2 - 2$ $p(x) = x^2 - 2$

$= pn(x)$

c $\dfrac{1}{4x + 12} = \dfrac{1}{2(2x + 4) + 4}$ $n(x) = 2x + 4$

$n(x) = 2x + 4$

$= \dfrac{1}{2n(x) + 4}$ $m(x) = \dfrac{1}{x}$

$= \dfrac{1}{nn(x)}$ $nn(x) = n^2(x)$

$= mnn(x)$

$= mn^2(x)$

Exercise 2D

1 Given the functions $f(x) = 4x + 1$, $g(x) = x^2 - 4$ and $h(x) = \dfrac{1}{x}$, find expressions for the functions:

a fg(x) **b** gf(x) **c** gh(x) **d** fh(x) **e** $f^2(x)$

2 For the following functions f(x) and g(x), find the composite functions fg(x) and gf(x). In each case find a suitable domain and the corresponding range when

a $f(x) = x - 1$, $g(x) = x^2$ **b** $f(x) = x - 3$, $g(x) = +\sqrt{x}$ **c** $f(x) = 2^x$, $g(x) = x + 3$

3 If $f(x) = 3x - 2$ and $g(x) = x^2$, find the number(s) a such that $fg(a) = gf(a)$.

4 Given that $s(x) = \dfrac{1}{x - 2}$ and $t(x) = 3x + 4$ find the number m such that $ts(m) = 16$.

5 The functions l(x), m(x), n(x) and p(x) are defined by $l(x) = 2x + 1$, $m(x) = x^2 - 1$, $n(x) = \dfrac{1}{x + 5}$ and $p(x) = x^3$. Find in terms of l, m, n and p the functions:

a $4x + 3$ **b** $4x^2 + 4x$ **c** $\dfrac{1}{x^2 + 4}$ **d** $\dfrac{2}{x + 5} + 1$

e $(x^2 - 1)^3$ **f** $2x^2 - 1$ **g** x^{27}

Functions

6 If $m(x) = 2x + 3$ and $n(x) = \dfrac{x - 3}{2}$, prove that $mn(x) = x$.

7 If $s(x) = \dfrac{3}{x + 1}$ and $t(x) = \dfrac{3 - x}{x}$, prove that $st(x) = x$.

8 If $f(x) = \dfrac{1}{x + 1}$, prove that $f^2(x) = \dfrac{x + 1}{x + 2}$. Hence find an expression for $f^3(x)$.

2.5 **The inverse function performs the opposite operation to the function. It takes elements of the range and maps them back into elements of the domain. For this reason inverse functions only exist for one-to-one functions.**

The inverse of $f(x)$ is written $f^{-1}(x)$

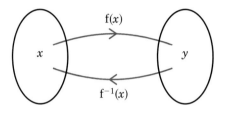

$ff^{-1}(x) = f^{-1}f(x) = x$

Function		Inverse		
$f(x) = x + 4$	'add 4'	'subtract 4'		$f^{-1}(x) = x - 4$
$g(x) = 5x$	'times 5'	'divide by 5'		$g^{-1}(x) = \dfrac{x}{5}$
$h(x) = 4x + 2$	'times 4, add 2'	'subtract 2, divide by 4'		$h^{-1}(x) = \dfrac{x - 2}{4}$

For many straightforward functions the inverse can be found using a flow chart.

Example 9

Find the inverse of the function $h(x) = 2x^2 - 7$.

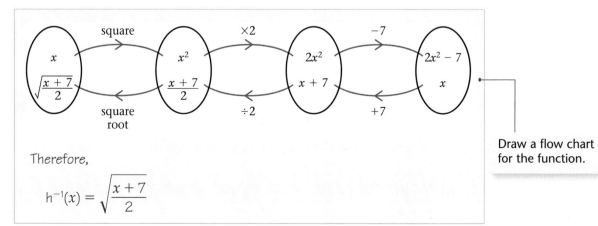

Draw a flow chart for the function.

Therefore,

$$h^{-1}(x) = \sqrt{\dfrac{x + 7}{2}}$$

23

Example 10

Find the inverse of the function $f(x) = \dfrac{3}{x-1}$, $\{x \in \mathbb{R}, x \neq 1\}$, by changing the subject of the formula.

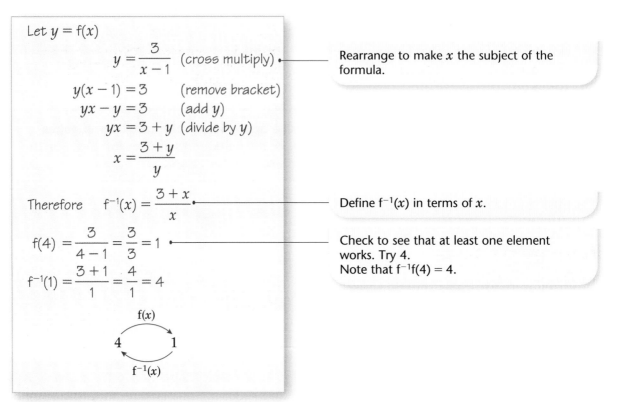

Let $y = f(x)$

$$y = \frac{3}{x-1} \quad \text{(cross multiply)}$$

Rearrange to make x the subject of the formula.

$$y(x-1) = 3 \quad \text{(remove bracket)}$$
$$yx - y = 3 \quad \text{(add } y)$$
$$yx = 3 + y \quad \text{(divide by } y)$$
$$x = \frac{3+y}{y}$$

Therefore $f^{-1}(x) = \dfrac{3+x}{x}$

Define $f^{-1}(x)$ in terms of x.

$$f(4) = \frac{3}{4-1} = \frac{3}{3} = 1$$
$$f^{-1}(1) = \frac{3+1}{1} = \frac{4}{1} = 4$$

Check to see that at least one element works. Try 4.
Note that $f^{-1}f(4) = 4$.

Example 11

The function $f(x)$ is defined by $f(x) = \sqrt{x-2}$ $\{x \in \mathbb{R}, x \geqslant 2\}$. Find the function $f^{-1}(x)$ in a similar form stating its domain.

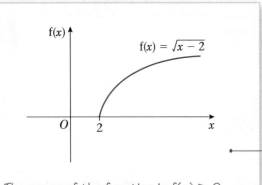

The range of the function is the domain of the inverse function and vice versa. This fact can be used to solve many problems where the domain is not all of $\mathbb{R}$.

Sketch the function for the values of x given.

The range of the function $=$ the domain of the inverse function.

The range of the function is $f(x) \geqslant 0$.
Therefore the domain of the inverse function is $x \geqslant 0$.

Change the subject of the formula.

$$y = \sqrt{x-2}$$
$$y^2 = x - 2$$
$$x = y^2 + 2$$

The inverse function is $f^{-1}(x) = x^2 + 2$ •────── Always write your function in terms of x.
$\{x \in \mathbb{R}, x \geqslant 0\}$

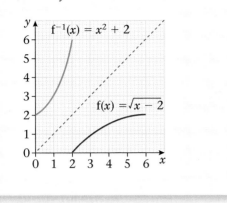

Note that the graph of $f^{-1}(x)$ is a reflection of $f(x)$ in the line $y = x$. This is because the reflection transforms y to x and x to y.

Example 12

If $g(x)$ is defined as $g(x) = 2x - 4$ $\{x \in \mathbb{R}, x \geqslant 0\}$

a Calculate $g^{-1}(x)$.

b Sketch the graphs of both functions on the same set of axes.

c What is the connection between the graphs?

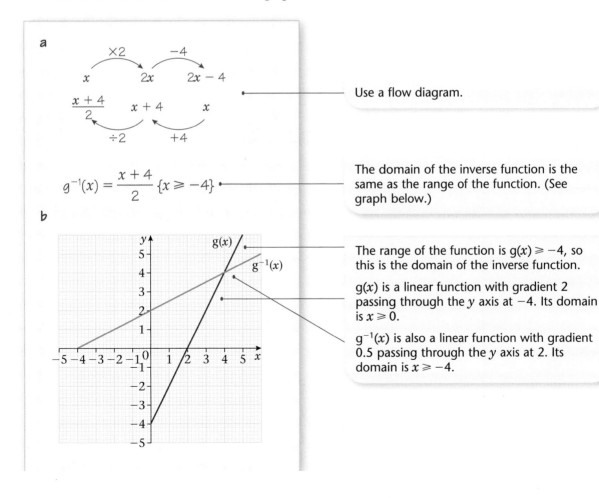

Use a flow diagram.

The domain of the inverse function is the same as the range of the function. (See graph below.)

The range of the function is $g(x) \geqslant -4$, so this is the domain of the inverse function.

$g(x)$ is a linear function with gradient 2 passing through the y axis at -4. Its domain is $x \geqslant 0$.

$g^{-1}(x)$ is also a linear function with gradient 0.5 passing through the y axis at 2. Its domain is $x \geqslant -4$.

c The graphs of $g^{-1}(x)$ and $g(x)$ are mirror images of each other in the line $y = x$.

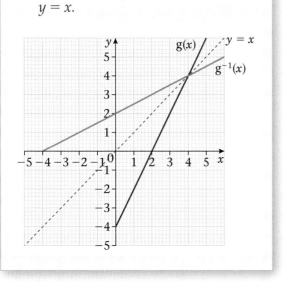

Example 13

The function $f(x)$ is defined by $f(x) = x^2 - 3$ $\{x \in \mathbb{R}, x \geqslant 0\}$.

a Find $f^{-1}(x)$ in similar terms.

b Sketch $f^{-1}(x)$.

c Find values of x such that $f(x) = f^{-1}(x)$.

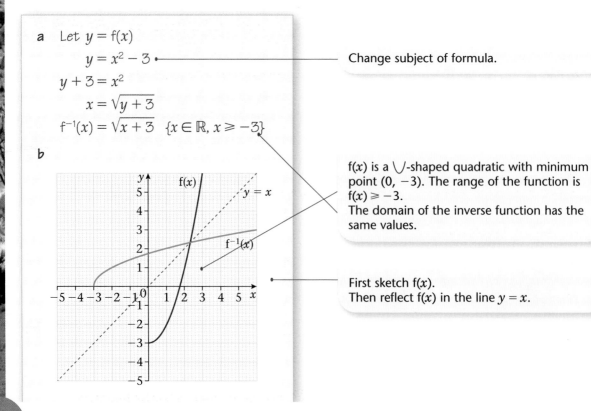

a Let $y = f(x)$

$y = x^2 - 3$ ———— Change subject of formula.

$y + 3 = x^2$

$x = \sqrt{y + 3}$

$f^{-1}(x) = \sqrt{x + 3}$ $\{x \in \mathbb{R}, x \geqslant -3\}$

b

$f(x)$ is a $\bigvee$-shaped quadratic with minimum point $(0, -3)$. The range of the function is $f(x) \geqslant -3$.
The domain of the inverse function has the same values.

First sketch $f(x)$.
Then reflect $f(x)$ in the line $y = x$.

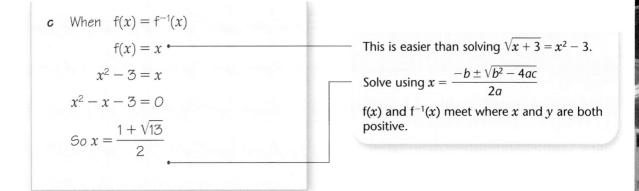

c When $f(x) = f^{-1}(x)$

$f(x) = x$ ————— This is easier than solving $\sqrt{x+3} = x^2 - 3$.

$x^2 - 3 = x$

$x^2 - x - 3 = 0$ ———— Solve using $x = \dfrac{-b \pm \sqrt{b^2 - 4ac}}{2a}$

So $x = \dfrac{1 + \sqrt{13}}{2}$

$f(x)$ and $f^{-1}(x)$ meet where x and y are both positive.

Exercise 2E

1 For the following functions $f(x)$, sketch the graphs of $f(x)$ and $f^{-1}(x)$ on the same set of axes. Determine also the equation of $f^{-1}(x)$.

a $f(x) = 2x + 3 \; \{x \in \mathbb{R}\}$

b $f(x) = \dfrac{x}{2} \; \{x \in \mathbb{R}\}$

c $f(x) = \dfrac{1}{x} \; \{x \in \mathbb{R}, \, x \neq 0\}$

d $f(x) = 4 - x \; \{x \in \mathbb{R}\}$

e $f(x) = x^2 + 2 \; \{x \in \mathbb{R}, \, x \geqslant 0\}$

f $f(x) = x^3 \; \{x \in \mathbb{R}\}$

2 Determine which of the functions in Question 1 are self inverses. (That is to say the function and its inverse are identical.)

3 Explain why the function $g(x) = 4 - x \; \{x \in \mathbb{R}, \, x > 0\}$ is not identical to its inverse.

4 For the following functions $g(x)$, sketch the graphs of $g(x)$ and $g^{-1}(x)$ on the same set of axes. Determine the equation of $g^{-1}(x)$, taking care with its domain.

a $g(x) = \dfrac{1}{x} \; \{x \in \mathbb{R}, \, x \geqslant 3\}$

b $g(x) = 2x - 1 \; \{x \in \mathbb{R}, \, x \geqslant 0\}$

c $g(x) = \dfrac{3}{x - 2} \; \{x \in \mathbb{R}, \, x > 2\}$

d $g(x) = \sqrt{x - 3} \; \{x \in \mathbb{R}, \, x \geqslant 7\}$

e $g(x) = x^2 + 2 \; \{x \in \mathbb{R}, \, x > 4\}$

f $g(x) = x^3 - 8 \; \{x \in \mathbb{R}, \, x \leqslant 2\}$

5 The function $m(x)$ is defined by $m(x) = x^2 + 4x + 9$ $\{x \in \mathbb{R}, \, x > a\}$ for some constant a. If $m^{-1}(x)$ exists, state the least value of a and hence determine the equation of $m^{-1}(x)$. State its domain.

> **Hint:** Completing the square helps in these types of questions.

6 Determine $t^{-1}(x)$ if the function $t(x)$ is defined by $t(x) = x^2 - 6x + 5 \; \{x \in \mathbb{R}, \, x \geqslant 5\}$.

7 The function $h(x)$ is defined by $h(x) = \dfrac{2x + 1}{x - 2} \; \{x \in \mathbb{R}, \, x \neq 2\}$.

a What happens to the function as x approaches 2?

b Find $h^{-1}(3)$.

c Find $h^{-1}(x)$, stating clearly its domain.

d Find the elements of the domain that get mapped to themselves by the function.

8 The function f(x) is defined by $f(x) = 2x^2 - 3$ $\{x \in \mathbb{R}, x < 0\}$. Determine

   a $f^{-1}(x)$ clearly stating its domain

   b the values of a for which $f(a) = f^{-1}(a)$.

Mixed exercise **2F**

1 Categorise the following as

   i not a function

   ii a one-to-one function

   iii a many-to-one function.

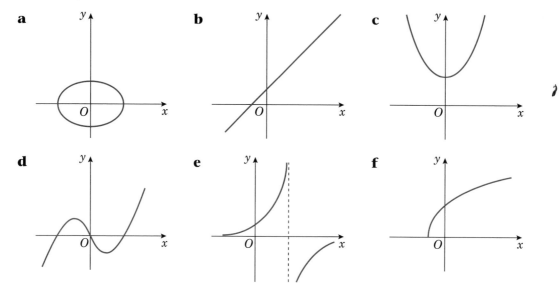

2 The following functions f(x), g(x) and h(x) are defined by

$$f(x) = 4(x - 2) \quad \{x \in \mathbb{R}, x \geqslant 0\}$$
$$g(x) = x^3 + 1 \quad \{x \in \mathbb{R}\}$$
$$h(x) = 3^x \quad \{x \in \mathbb{R}\}$$

   a Find $f(7)$, $g(3)$ and $h(-2)$.

   b Find the range of $f(x)$ and the range of $g(x)$.

   c Find $g^{-1}(x)$.

   d Find the composite function $fg(x)$.

   e Solve $gh(a) = 244$.

3 The function n(x) is defined by

$$n(x) = \begin{cases} 5 - x & x \leqslant 0 \\ x^2 & x > 0 \end{cases}$$

   a Find $n(-3)$ and $n(3)$.

   b Find the value(s) of a such that $n(a) = 50$.

4 The function g(x) is defined as $g(x) = 2x + 7$ $\{x \in \mathbb{R}, x \geqslant 0\}$.

   a Sketch $g(x)$ and find the range.

   b Determine $g^{-1}(x)$, stating its domain.

   c Sketch $g^{-1}(x)$ on the same axes as $g(x)$, stating the relationship between the two graphs.

5 The functions f and g are defined by

$$f : x \rightarrow 4x - 1 \; \{x \in \mathbb{R}\}$$
$$g : x \rightarrow \frac{3}{2x - 1} \; \{x \in \mathbb{R}, \, x \neq \tfrac{1}{2}\}$$

Find in its simplest form:

a the inverse function f^{-1}

b the composite function gf, stating its domain

c the values of x for which $2f(x) = g(x)$, giving your answers to 3 decimal places. **E**

6 The function $f(x)$ is defined by

$$f(x) = \begin{cases} -x & x \leq 1 \\ x - 2 & x > 1 \end{cases}$$

a Sketch the graph of $f(x)$ for $-2 \leq x \leq 6$.

b Find the values of x for which $f(x) = -\tfrac{1}{2}$. **E**

7 The function f is defined by

$$f : x \rightarrow \frac{2x + 3}{x - 1} \quad \{x \in \mathbb{R}, \, x > 1\}$$

a Find $f^{-1}(x)$.

b Find **i** the range of $f^{-1}(x)$
             ii the domain of $f^{-1}(x)$. **E**

8 The functions f and g are defined by

$$f : x \rightarrow \frac{x}{x - 2} \quad \{x \in \mathbb{R}, \, x \neq 2\}$$
$$g : x \rightarrow \frac{3}{x} \quad \{x \in \mathbb{R}, \, x \neq 0\}$$

a Find an expression for $f^{-1}(x)$.

b Write down the range of $f^{-1}(x)$.

c Calculate gf(1.5).

d Use algebra to find the values of x for which $g(x) = f(x) + 4$. **E**

9 The functions f and g are given by

$$f : x \rightarrow \frac{x}{x^2 - 1} - \frac{1}{x + 1} \quad \{x \in \mathbb{R}, \, x > 1\}$$
$$g : x \rightarrow \frac{2}{x} \quad \{x \in \mathbb{R}, \, x > 0\}$$

a Show that $f(x) = \dfrac{1}{(x - 1)(x + 1)}$.

b Find the range of $f(x)$.

c Solve gf(x) = 70. **E**

Summary of key points

1 A function is a special mapping such that every element of the domain is mapped to exactly one element in the range.

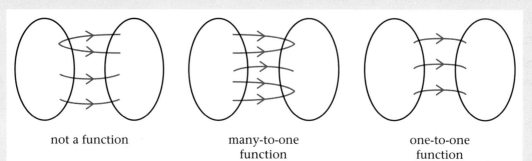

not a function   many-to-one function   one-to-one function

2 A one-to-one function is a special function where every element of the range has been mapped from exactly one element of the domain.

3 Many mappings can be made into functions by changing the domain. For example, the mapping 'positive square root' can be changed into the function $f(x) = \sqrt{x}$ by having a domain of $x \geqslant 0$.

4 If we combine two or more functions we can create a composite function. The function below is written $fg(x)$ as g acts on x first, then f acts on the result. For example,

$g(x) = 2x + 3, f(x) = x^2$
$fg(4) = f(2 \times 4 + 3) = f(11) = 11^2 = 121$

Similarly

$fg(x) = (2x + 3)^2$

5 The inverse of a function $f(x)$ is written $f^{-1}(x)$ and performs the opposite operation(s) to the function. To calculate the inverse function you change the subject of the formula. For example, the inverse function of $g(x) = 4x - 3$ is

$g^{-1}(x) = \dfrac{x + 3}{4}$

6 The range of the function is the domain of the inverse function and vice versa.

7 The graph of $f^{-1}(x)$ is a reflection of $f(x)$ in the line $y = x$.

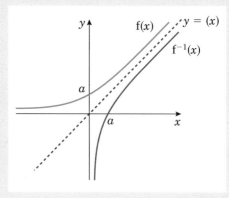

After completing this chapter you should be able to

1 sketch simple transformations of the graph of $y = e^x$

2 sketch simple transformations of the graph $y = \ln x$

3 solve equations involving e^x and $\ln x$

4 know what is meant by the terms exponential growth and decay

5 solve real life examples of exponential growth and decay.

The exponential and log functions

Exponential functions occur naturally in real life. Scientists can model the number of elephants in a herd by using an exponential function.

3.1 Exponential functions are ones of the form $y = a^x$. **Graphs of these functions all pass through (0, 1) because $a^0 = 1$ for any number a.**

Example 1

Sketch the graph of $f(x) = 2^x$ for the domain $x \in \mathbb{R}$. State the range of the function.

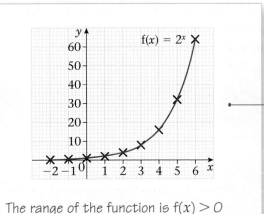

The range of the function is $f(x) > 0$

Draw up a table of values.

x	-2	-1	0	1	2	3	4	5	6
y	0.25	0.5	1	2	4	8	16	32	64

Plot points on a graph.

The gradient functions of these graphs are similar to the functions themselves.

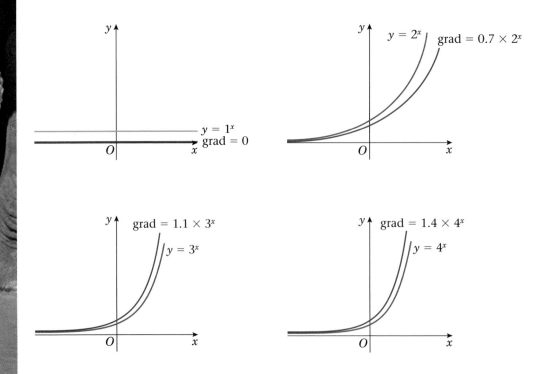

Put these results in a table.

Function	Gradient
$y = 1^x$	grad $= 0 \times 1^x$
$y = 2^x$	grad $= 0.7 \times 2^x$
$y = 3^x$	grad $= 1.1 \times 3^x$
$y = 4^x$	grad $= 1.4 \times 4^x$

You should be able to spot from this table that as a increases for the function $y = a^x$, so does the gradient function.

You should be able to deduce that there is going to be a number between 2 and 3 such that the gradient function would be the same as the function. This number is approximately equal to 2.718 and is represented by the letter 'e'. It is similar to π in the respect that it is an irrational number representing a number that exists in the real world.

■ **The exponential function $y = e^x$ (where e $\approx$ 2.718) is therefore the function in which the gradient is identical to the function. For this reason it is often referred to as the exponential function.**

> If $y = e^x$ then $\dfrac{dy}{dx} = e^x$

3.2 **All exponential graphs will follow a similar pattern. The standard graph of $y = e^x$ can be used to represent 'exponential' growth, which is how population growth can be modelled in real life.**

Example 2

Draw the graphs of:

a $y = e^x$ **b** $y = e^{-x}$

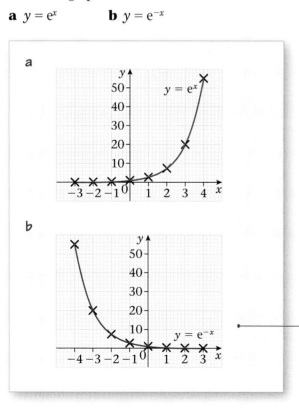

a

b

A table of values will show you how rapidly this curve grows.

x	-2	-1	0	1	2	3	4	5
y	0.14	0.37	1	2.7	7.4	20	55	148

With these curves it is worth keeping in mind:

- as $x \to \infty$, $e^x \to \infty$ (it grows very rapidly)
- when $x = 0$, $e^0 = 1$ [(0, 1) lies on the curve]
- as $x \to -\infty$, $e^x \to 0$ (it approaches but never reaches the x-axis).

This curve is similar to the one in part **a** except that its value at $x = 2$ is e^{-2} and its value at $x = -2$ is e^2.

Hence it is a reflection of the curve of part **a** in the y-axis.

The graph in Example **2b** is often referred to as exponential decay. It is used as a model in many examples from real life including the fall in value of a car as well as the decay in radioactive isotopes.

Example 3

Draw graphs of the exponential functions:

a $y = e^{2x}$ **b** $y = 10e^{-x}$ **c** $y = 3 + 4e^{\frac{1}{2}x}$

a $y = e^{2x}$
$\quad = (e^x)^2$

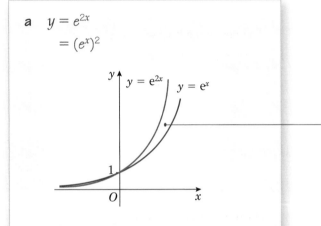

x	-3	0	3
y	0.002	1	403

If you calculate some values it can give you an idea of the shape of the graph.

The y values of $y = e^{2x}$ are the 'square' of the y values of $y = e^x$.

b $y = 10e^{-x}$

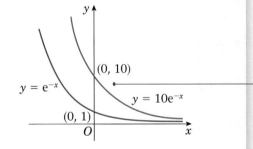

x	-3	0	3
y	201	10	0.5

Calculating some y values helps you sketch the curve.

The y values of $y = 10e^{-x}$ are 10 times bigger than the y values of $y = e^{-x}$.

c $y = 3 + 4e^{\frac{1}{2}x}$

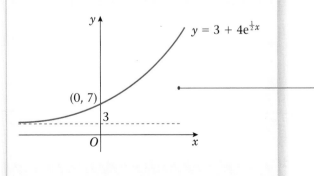

x	-3	0	3
y	3.9	7	21

Since $e^{\frac{1}{2}x} > 0$

$3 + 4e^{\frac{1}{2}x} > 3$.
Range of function is $y > 3$.

Example 4

The price of a used car can be represented by the formula

$$P = 16\,000\,e^{-\frac{t}{10}}$$

where P is the price in £'s and t is the age in years from new.

Calculate:

a the new price

b the value at 5 years old

c what the model suggests about the eventual value of the car.

Use this to sketch the graph of P against t.

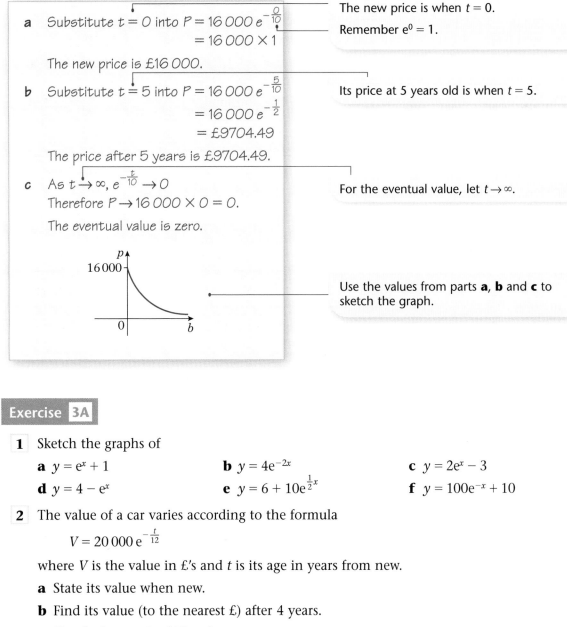

a Substitute $t = 0$ into $P = 16\,000\,e^{-\frac{0}{10}}$

$\qquad = 16\,000 \times 1$

The new price is £16 000.

The new price is when $t = 0$.

Remember $e^0 = 1$.

b Substitute $t = 5$ into $P = 16\,000\,e^{-\frac{5}{10}}$

$\qquad = 16\,000\,e^{-\frac{1}{2}}$

$\qquad = £9704.49$

The price after 5 years is £9704.49.

Its price at 5 years old is when $t = 5$.

c As $t \to \infty$, $e^{-\frac{t}{10}} \to 0$

Therefore $P \to 16\,000 \times 0 = 0$.

The eventual value is zero.

For the eventual value, let $t \to \infty$.

Use the values from parts **a**, **b** and **c** to sketch the graph.

Exercise 3A

1 Sketch the graphs of

a $y = e^x + 1$
  b $y = 4e^{-2x}$
  c $y = 2e^x - 3$

d $y = 4 - e^x$
  e $y = 6 + 10e^{\frac{1}{2}x}$
  f $y = 100e^{-x} + 10$

2 The value of a car varies according to the formula

$$V = 20\,000\,e^{-\frac{t}{12}}$$

where V is the value in £'s and t is its age in years from new.

a State its value when new.

b Find its value (to the nearest £) after 4 years.

c Sketch the graph of V against t.

3 The population of a country is increasing according to the formula

$$P = 20 + 10\,e^{\frac{t}{50}}$$

where P is the population in thousands and t is the time in years after the year 2000.

a State the population in the year 2000.

b Use the model to predict the population in the year 2020.

c Sketch the graph of P against t for the years 2000 to 2100.

4 The number of people infected with a disease varies according to the formula

$$N = 300 - 100\,e^{-0.5t}$$

where N is the number of people infected with the disease and t is the time in years after detection.

a How many people were first diagnosed with the disease?

b What is the long term prediction of how this disease will spread?

c Graph N against t.

5 The value of an investment varies according to the formula

$$V = A\,e^{\frac{t}{12}}$$

where V is the value of the investment in £'s, A is a constant to be found and t is the time in years after the investment was made.

a If the investment was worth £8000 after 3 years find A to the nearest £.

b Find the value of the investment after 10 years.

c By what factor will be the original investment have increased by after 20 years?

3.3 To study the exponential function further, it becomes necessary to introduce its inverse function. From Chapter 2 you should know that inverse functions perform the 'opposite' operation to a function, in exactly the same way as '+4' and '−4' and 'x^2' and '$\sqrt{x}$' are inverse operations.

■ the inverse to e^x is $\log_e x$ (often written $\ln x$).

Example 5

Solve the equations

a $e^x = 3$ **b** $\ln x = 4$

a When $e^x = 3$
$x = \ln 3$

b When $\ln x = 4$
$x = e^4$

The key to solving any equation is knowing the inverse operation.
When $x^2 = 10$, $x = \sqrt{10}$.

The inverse of e^x is $\ln x$ and vice versa.

Using your knowledge of inverse functions, the graph of $\ln x$ will be a reflection of e^x in the line $y = x$.

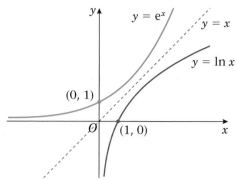

The function $f(x) = \ln x$ therefore has a domain of $\{x \in \mathbb{R}, x > 0\}$ and a range of $\{f(x) \in \mathbb{R}\}$.

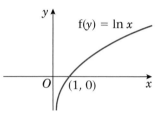

The important points about the graph $y = \ln x$ are:

- as $x \to 0$, $y \to -\infty$
- $\ln x$ does not exist for negative numbers
- when $x = 1$, $y = 0$
- as $x \to \infty$, $y \to \infty$ (slowly).

Example 6

Sketch the graphs of

a $y = \ln(3 - x)$ **b** $y = 3 + \ln(2x)$

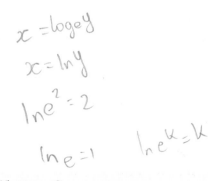

$x = \log_e y$

$x = \ln y$

$\ln e^2 = 2$

$\ln e = 1$ $\ln e^k = k$

a $y = \ln(3 - x)$

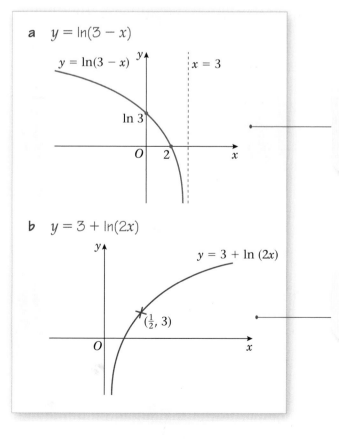

When $x \to 3$, $y \to -\infty$.

y does not exist for values of x bigger than 3.

When $x = 2$, $y = \ln(3 - 2) = \ln 1 = 0$.

As $x \to -\infty$, $y \to \infty$ (slowly).

b $y = 3 + \ln(2x)$

When $x \to 0$, $y \to -\infty$.

When $x = \frac{1}{2}$, $y = 3 + \ln 1 = 3$.

As $x \to \infty$, $y \to \infty$ (slowly).

Example 7

Solve the equations:

a $e^{2x+3} = 4$ **b** $2\ln x + 1 = 5$

> These questions are solved by changing the subject of the formula and using the fact that $\ln x$ and e^x are inverse functions.

a $e^{2x+3} = 4$

$2x + 3 = \ln 4$

$2x = \ln 4 - 3$

$x = \dfrac{\ln 4 - 3}{2}$

$ = \dfrac{\ln 4}{2} - \dfrac{3}{2}$

$ = \ln 2 - \dfrac{3}{2}$

> The inverse to e^x is $\ln x$.

> Sometimes questions ask you to put an answer in a particular form. Note that
> $\dfrac{\ln 4}{2} = \tfrac{1}{2}\ln 4 = \ln 4^{\frac{1}{2}} = \ln 2$

b $2\ln x + 1 = 5$

$2\ln x = 4$

$\ln x = 2$

$x = e^2$

> Isolate $\ln x$.
> The inverse of $\ln x$ is e^x.

Example 8

The number of elephants in a herd can be represented by the equation

$$N = 150 - 80\,e^{-\frac{t}{40}}$$

where N is the number of elephants in the herd and t is the time in years after the year 2003.

Calculate:

a the number of elephants in the herd in 2003

b the number of elephants in the herd in 2007

c the year when the population will grow to above 100

d the long term population of the herd as predicted by the model.

Use all of the above information to sketch a graph of N against t for the model.

$N = 150 - 80\,e^{-\frac{t}{40}}$

a In the year 2003, $N = 150 - 80\,e^{-\frac{0}{40}}$

$ = 150 - 80 \times 1$

$ = 70$

> For 2003 substitute $t = 0$.
> $e^0 = 1$

b In the year 2007,

$N = 150 - 80\,e^{-\frac{4}{40}}$

$ = 150 - 72.4$

$ = 78$ (to nearest elephant)

> For 2007 substitute $t = 4$.

c If population is 100, then

$$100 = 150 - 80\,e^{-\frac{t}{40}}$$

$$80\,e^{-\frac{t}{40}} = 50$$

$$e^{-\frac{t}{40}} = \frac{50}{80}$$

$$-\frac{t}{40} = \ln\frac{50}{80}$$

$$t = 18.8 \text{ years}$$

Therefore the population will be over 100 by the year 2022.

Substitute $N = 100$.

Isolate $e^{-\frac{t}{40}}$.

The inverse of e^x is $\ln x$.

Add 18.8 years to 2003 and round up answer.

d As $t \to \infty$, $e^{-\frac{t}{40}} \to 0$ and therefore

$$N \to 150 - 80 \times 0 = 150$$

The long term population as predicted by the model is 150.

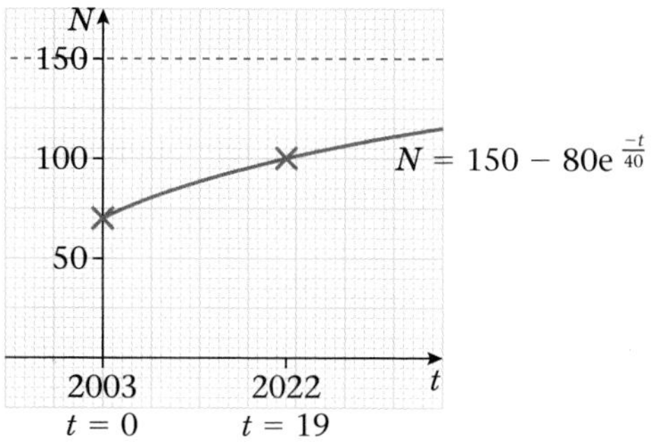

Long term suggests as $t \to \infty$.

$N = 150 - 80e^{\frac{-t}{40}}$

Use all the above information to sketch the graph.

Exercise 3B

1 Solve the following equations giving exact solutions:

a $e^x = 5$
b $\ln x = 4$
c $e^{2x} = 7$
d $\ln\frac{x}{2} = 4$
e $e^{x-1} = 8$
f $\ln(2x + 1) = 5$
g $e^{-x} = 10$
h $\ln(2 - x) = 4$
i $2e^{4x} - 3 = 8$

2 Solve the following giving your solution in terms of $\ln 2$:

a $e^{3x} = 8$
b $e^{-2x} = 4$
c $e^{2x+1} = 0.5$

3 Sketch the following graphs stating any asymptotes and intersections with axes:

a $y = \ln(x + 1)$
b $y = 2\ln x$
c $y = \ln(2x)$
d $y = (\ln x)^2$
e $y = \ln(4 - x)$
f $y = 3 + \ln(x + 2)$

4 The price of a new car varies according to the formula

$$P = 15\,000\,e^{-\frac{t}{10}}$$

where P is the price in £'s and t is the age in years from new.

a State its new value.

b Calculate its value after 5 years (to the nearest £).

c Find its age when its price falls below £5000.

d Sketch the graph showing how the price varies over time. Is this a good model?

5 The graph opposite is of the function
$f(x) = \ln(2 + 3x)$ $\{x \in \mathbb{R}, x > a\}$.

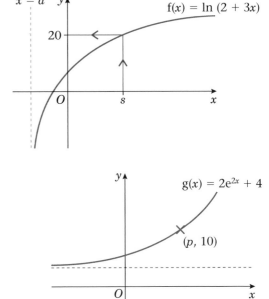

a State the value of a.

b Find the value of s for which $f(s) = 20$.

c Find the function $f^{-1}(x)$ stating its domain.

d Sketch the graphs $f(x)$ and $f^{-1}(x)$ on the same axes stating the relationship between them.

6 The graph opposite is of the function
$g(x) = 2e^{2x} + 4$ $\{x \in \mathbb{R}\}$.

a Find the range of the function.

b Find the value of p to 2 significant figures.

c Find $g^{-1}(x)$ stating its domain.

d Sketch $g(x)$ and $g^{-1}(x)$ on the same set of axes stating the relationship between them.

7 The number of bacteria in a culture grows according to the following equation:

$$N = 100 + 50\,e^{\frac{t}{30}}$$

where N is the number of bacteria present and t is the time in days from the start of the experiment.

a State the number of bacteria present at the start of the experiment.

b State the number after 10 days.

c State the day on which the number first reaches $1\,000\,000$.

d Sketch the graph showing how N varies with t.

8 The graph opposite shows the function
$h(x) = 40 - 10\,e^{3x}$ $\{x > 0, x \in \mathbb{R}\}$.

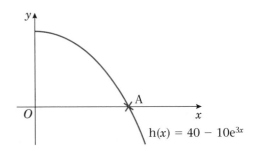

a State the range of the function.

b Find the exact coordinates of A in terms of $\ln 2$.

c Find $h^{-1}(x)$ stating its domain.

Mixed exercise 3C

1 Sketch the following functions stating any asymptotes and intersections with axes:

a $y = e^x + 3$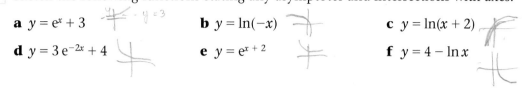

b $y = \ln(-x)$

c $y = \ln(x + 2)$

d $y = 3\,e^{-2x} + 4$

e $y = e^{x+2}$

f $y = 4 - \ln x$

2 Solve the following equations, giving exact solutions:

a $\ln(2x - 5) = 8$

$2x - 5 = e^8 \quad x = \dfrac{e^8 + 5}{2}$

b $e^{4x} = 5$ $\quad 4x = e^5$

c $24 - e^{-2x} = 10$ $\quad \ln 14 = -2x$

d $\ln x + \ln(x - 3) = 0$

$\ln(x + x - 3) = 0$

$2x - 3 = 1 \quad x = 2$

e $e^x + e^{-x} = 2$

f $\ln 2 + \ln x = 4$

3 The function $c(x) = 3 + \ln(4 - x)$ is shown below.

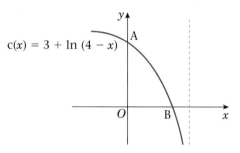

a State the exact coordinates of point A. $\quad 3 + \ln 4$

b Calculate the exact coordinates of point B.

c Find the inverse function $c^{-1}(x)$ stating its domain.

d Sketch $c(x)$ and $c^{-1}(x)$ on the same set of axes stating the relationship between them.

4 The price of a computer system can be modelled by the formula

$$P = 100 + 850\,e^{-\frac{t}{2}}$$

where P is the price of the system in £s and t is the age of the computer in years after being purchased.

a Calculate the new price of the system.

b Calculate its price after 3 years.

c When will it be worth less than £200?

d Find its price as $t \to \infty$.

e Sketch the graph showing P against t.

Comment on the appropriateness of this model.

5 The function f is defined by

$$f:x \rightarrow \ln(5x - 2) \ \{x \in \mathbb{R}, \ x > \tfrac{2}{5}\}.$$

a Find an expression for $f^{-1}(x)$.

b Write down the domain of $f^{-1}(x)$.

c Solve, giving your answer to 3 decimal places,

$$\ln(5x - 2) = 2.$$ **E**

6 The functions f and g are given by

$$f:x \rightarrow 3x - 1 \ \{x \in \mathbb{R}\}$$
$$g:x \rightarrow e^{\frac{x}{2}} \ \{x \in \mathbb{R}\}$$

a Find the value of fg(4), giving your answer to 2 decimal places.

b Express the inverse function $f^{-1}(x)$ in the form $f^{-1}:x \rightarrow \dots$.

c Using the same axes, sketch the graphs of the functions f and gf. Write on your sketch the value of each function at $x = 0$.

d Find the values of x for which $f^{-1}(x) = \dfrac{5}{f(x)}$. **E**

7 The points P and Q lie on the curve with equation $y = e^{\frac{1}{2}x}$. The x-coordinates of P and Q are ln 4 and ln 16 respectively.

a Find an equation for the line PQ.

b Show that this line passes through the origin O.

c Calculate the length, to 3 significant figures, of the line segment PQ. **E**

8 The functions f and g are defined over the set of real numbers by

$$f:x \rightarrow 3x - 5$$
$$g:x \rightarrow e^{-2x}$$

a State the range of g(x).

b Sketch the graphs of the inverse functions f^{-1} and g^{-1} and write on your sketches the coordinates of any points at which a graph meets the coordinate axes.

c State, giving a reason, the number of roots of the equation

$$f^{-1}(x) = g^{-1}(x).$$

d Evaluate $fg(-\tfrac{1}{3})$, giving your answer to 2 decimal places.

9 The function f is defined by $f:x \rightarrow e^x + k$, $x \in \mathbb{R}$ and k is a positive constant.

a State the range of f(x).

b Find f(ln k), simplifying your answer.

c Find f^{-1}, the inverse function of f, in the form $f^{-1}:x \rightarrow \dots$, stating its domain.

d On the same axes, sketch the curves with equations $y = f(x)$ and $y = f^{-1}(x)$, giving the coordinates of all points where the graphs cut the axes. **E**

10 The function f is given by

$$f : x \to \ln(4 - 2x) \quad \{x \in \mathbb{R}, \, x < 2\}$$

a Find an expression for $f^{-1}(x)$.

b Sketch the curve with equation $y = f^{-1}(x)$, showing the coordinates of the points where the curve meets the axes.

c State the range of $f^{-1}(x)$.

The function g is given by

$$g : x \to e^x \quad \{x \in \mathbb{R}\}$$

d Find the value of gf(0.5).

         E

11 The function $f(x)$ is defined by

$$f(x) = 3x^3 - 4x^2 - 5x + 2$$

a Show that $(x + 1)$ is a factor of $f(x)$.

b Factorise $f(x)$ completely.

c Solve, giving your answers to 2 decimal places, the equation

$$3[\ln(2x)]^3 - 4[\ln(2x)]^2 - 5\ln(2x) + 2 = 0 \quad x > 0$$

         E

Summary of key points

1 Exponential functions are ones of the form $y = a^x$. They all pass through the point (0, 1).

   The domain is all the real numbers. The range is $f(x) > 0$.

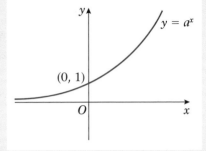

2 The exponential function $y = e^x$ (where $e \approx 2.718$) is a special function whose gradient is identical to the function.

3 The inverse function to e^x is $\ln x$.

4 The natural log function is a reflection of $y = e^x$ in the line $y = x$. It passes through the point (1, 0).

   The domain is the positive numbers. The range is all the real numbers.

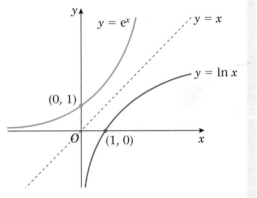

5 To solve an equation using $\ln x$ or e^x you must change the subject of the formula and use the fact that they are inverses of each other.

6 Growth and decay models are based around the exponential equations

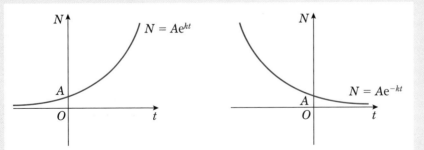

where A and k are positive numbers.

After completing this chapter you should be able to

1 use a graphical method to find the number of roots of the equation $f(x) = 0$

2 prove that a root lies within a given interval $[a, b]$

3 use iteration to find an approximation to the root of the equation $f(x) = 0$

4 express your answer to an appropriate degree of accuracy.

Numerical methods

The branch of Mathematics called Numerical Analysis predates the invention of modern computers by many centuries. Many equations don't have exact solutions and iterative methods form successive approximations that *converge* to the exact solution.

Weather forecasters use numerical methods to predict storms.

> **4.1** You can find approximations for the roots of the equation f(x) = 0 graphically.

Example **1**

Show that the equation $x^3 - 3x^2 + 3x - 4 = 0$ has a root between $x = 2$ and $x = 3$.

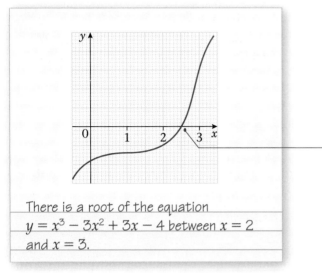

There is a root of the equation
$y = x^3 - 3x^2 + 3x - 4$ between $x = 2$
and $x = 3$.

Draw $y = x^3 - 3x^2 + 3x - 4$.

Find the point on the curve
$y = x^3 - 3x^2 + 3x - 4$ at which $y = 0$.

Remember the line $y = 0$ is the x-axis. So find where $y = x^3 - 3x^2 + 3x - 4$ crosses the x-axis.

The graph crosses the x-axis between $x = 2$ and $x = 3$, so there is a root of the equation between $x = 2$ and $x = 3$.

Example **2**

Show that the values of y for points on the graph of $y = 4 + 2x - x^3$ change sign as the graph crosses the x-axis.

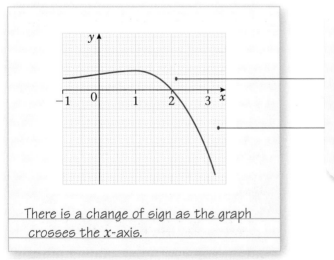

There is a change of sign as the graph
crosses the x-axis.

Draw $y = 4 + 2x - x^3$.

The graph meets the x-axis at $x = 2$.

To the left of $x = 2$ the values of y are positive because the curve lies above the x-axis at these points.

To the right of $x = 2$ the values of y are negative because the curve lies below the x-axis at these points.

Example 3

Show that $e^x + 2x - 3 = 0$ has a root between $x = 0.5$ and $x = 0.6$.

Let $f(x) \equiv e^x + 2x - 3$

$f(0.5) = e^{0.5} + 2(0.5) - 3$

$\quad\quad = 1.648\ldots + 1 - 3$

$\quad\quad = -0.351\ldots$

$f(0.6) = e^{0.6} + 2(0.6) - 3$

$\quad\quad = 1.822\ldots + 1.2 - 3$

$\quad\quad = 0.022\ldots$

There is a root between $x = 0.5$ and $x = 0.6$.

Show that the graph of $y = f(x)$ crosses the x-axis between $x = 0.5$ and $x = 0.6$.

Substitute $x = 0.5$ and $x = 0.6$ into the function.

$f(0.5) < 0$ and $f(0.6) > 0$, so there is a change of sign.

The graph of $y = f(x)$ crosses the x-axis between $x = 0.5$ and $x = 0.6$, so there is a root between $x = 0.5$ and $x = 0.6$.

■ **In general, if you find an interval in which $f(x)$ changes sign, then the interval must contain a root of the equation $f(x) = 0$.**

The only exception to this is when $f(x)$ has a discontinuity in the interval, e.g. $f(x) = \dfrac{1}{x}$ has a discontinuity at $x = 0$.

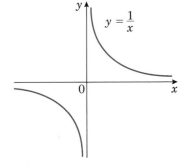

The graph shows that to the left of $x = 0$, $f(x) < 0$, and to the right of $x = 0$, $f(x) > 0$.
So the function changes sign in any interval that contains $x = 0$ but $x = 0$ is not a root of the equation $f(x) = 0$. There is a discontinuity at $x = 0$.

Example 4

a Using the same axes, sketch the graphs of $y = \ln x$ and $y = \dfrac{1}{x}$. Hence show that the equation $\ln x = \dfrac{1}{x}$ has only one root.

b Show that this root lies in the interval $1.7 < x < 1.8$.

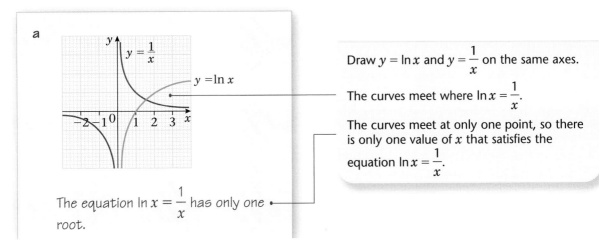

a

The equation $\ln x = \dfrac{1}{x}$ has only one root.

Draw $y = \ln x$ and $y = \dfrac{1}{x}$ on the same axes.

The curves meet where $\ln x = \dfrac{1}{x}$.

The curves meet at only one point, so there is only one value of x that satisfies the equation $\ln x = \dfrac{1}{x}$.

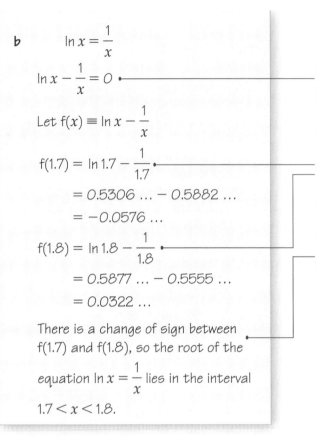

b

$$\ln x = \frac{1}{x}$$

$$\ln x - \frac{1}{x} = 0$$ ●————————— Rearrange the equation into the form f(x) = 0.

Subtract $\frac{1}{x}$ from each side.

Let $f(x) \equiv \ln x - \frac{1}{x}$

Show that f(x) = 0 has a root between $x = 1.7$ and $x = 1.8$.

$$f(1.7) = \ln 1.7 - \frac{1}{1.7}$$ ●

Substitute $x = 1.7$ and $x = 1.8$ into the function.

$$= 0.5306 \ldots - 0.5882 \ldots$$

$$= -0.0576 \ldots$$

f(1.7) < 0 and f(1.8) > 0, so there is a change of sign.

$$f(1.8) = \ln 1.8 - \frac{1}{1.8}$$ ●

The graph of $y = f(x)$ crosses the x-axis between $x = 1.7$ and $x = 1.8$, so there is a root between $x = 1.7$ and $x = 1.8$.

$$= 0.5877 \ldots - 0.5555 \ldots$$

$$= 0.0322 \ldots$$

There is a change of sign between f(1.7) and f(1.8), so the root of the

equation $\ln x = \frac{1}{x}$ lies in the interval

$$1.7 < x < 1.8.$$

Exercise 4A

1 Show that each of these equations f(x) = 0 has a root in the given interval(s):

  a $x^3 - x + 5 = 0$ $-2 < x < -1$. $-8 + 2 + 5 = -$ $-1 + 1 + 5 = +$

  b $3 + x^2 - x^3 = 0$ $1 < x < 2$.

  c $x^2 - \sqrt{x} - 10 = 0$ $3 < x < 4$.

  d $x^3 - \frac{1}{x} - 2 = 0$ $-0.5 < x < -0.2$ and $1 < x < 2$.

  e $x^5 - 5x^3 - 10 = 0$ $-2 < x < -1.8$, $-1.8 < x < -1$ and $2 < x < 3$.

  f $\sin x - \ln x = 0$ $2.2 < x < 2.3$

  g $e^x - \ln x - 5 = 0$ $1.65 < x < 1.75$.

  h $\sqrt[3]{x} - \cos x = 0$ $0.5 < x < 0.6$.

> For parts **f** and **h**, remember to use radians.

2 Given that $f(x) = x^3 - 5x^2 + 2$, show that the equation f(x) = 0 has a root near to $x = 5$.

  when $x = 4$ $f(x) = -$ when $x = 5$ $f(x) = +$

3 Given that $f(x) \equiv 3 - 5x + x^3$, show that the equation f(x) = 0 has a root $x = a$, where a lies in the interval $1 < a < 2$. $f(1) = -$ $f(2) = +$

4 Given that $f(x) \equiv e^x \sin x - 1$, show that the equation f(x) = 0 has a root $x = r$, where r lies in the interval $0.5 < r < 0.6$. $f(0.51) = -0.014$ $f(0.53) = -2.014$

5 It is given that $f(x) \equiv x^3 - 7x + 5$.

  a Copy and complete the table below.

x	-3	-2	-1	0	1	2	3
$f(x)$	-1	11					

  b Given that the negative root of the equation $x^3 - 7x + 5 = 0$ lies between α and $\alpha + 1$, where α is an integer, write down the value of α.

6 Given that $f(x) \equiv x - (\sin x + \cos x)^{\frac{1}{2}}$, $0 \leqslant x \leqslant \frac{3}{4}\pi$, show that the equation $f(x) = 0$ has a root lying between $\dfrac{\pi}{3}$ and $\dfrac{\pi}{2}$.

7 **a** Using the same axes, sketch the graphs of $y = e^{-x}$ and $y = x^2$.

  b Explain why the equation $e^{-x} = x^2$ has only one root.

  c Show that the equation $e^{-x} = x^2$ has a root between $x = 0.70$ and $x = 0.71$.

8 **a** On the same axes, sketch the graphs of $y = \ln x$ and $y = e^x - 4$.

  b Write down the number of roots of the equation $\ln x = e^x - 4$.

  c Show that the equation $\ln x = e^x - 4$ has a root in the interval $(1.4, 1.5)$.

9 **a** On the same axes, sketch the graphs of $y = \sqrt{x}$ and $y = \dfrac{2}{x}$.

  b Using your sketch, write down the number of roots of the equation $\sqrt{x} = \dfrac{2}{x}$.

  c Given that $f(x) \equiv \sqrt{x} - \dfrac{2}{x}$, show that $f(x) = 0$ has a root r, where r lies between $x = 1$ and $x = 2$.

  d Show that the equation $\sqrt{x} = \dfrac{2}{x}$ may be written in the form $x^p = q$, where p and q are integers to be found.

  e Hence write down the exact value of the root of the equation $\sqrt{x} - \dfrac{2}{x} = 0$.

10 **a** On the same axes, sketch the graphs of $y = \dfrac{1}{x}$ and $y = x + 3$.

  b Write down the number of roots of the equation $\dfrac{1}{x} = x + 3$.

  c Show that the positive root of the equation $\dfrac{1}{x} = x + 3$ lies in the interval $(0.30, 0.31)$.

  d Show that the equation $\dfrac{1}{x} = x + 3$ may be written in the form $x^2 + 3x - 1 = 0$.

  e Use the quadratic formula to find the positive root of the equation $x^2 + 3x - 1 = 0$ to 3 decimal places.

4.2 You can use iteration to find an approximation for the root of the equation f(x) = 0.

Example 5

a Show that $x^2 - 4x + 1 = 0$ can be written in the form $x = 4 - \dfrac{1}{x}$.

b Use the iteration formula $x_{n+1} = 4 - \dfrac{1}{x_n}$ to find, to 2 decimal places, a root of the equation $x^2 - 4x + 1 = 0$. Start with $x_0 = 3$.

c Show graphically the first two iterations of this formula.

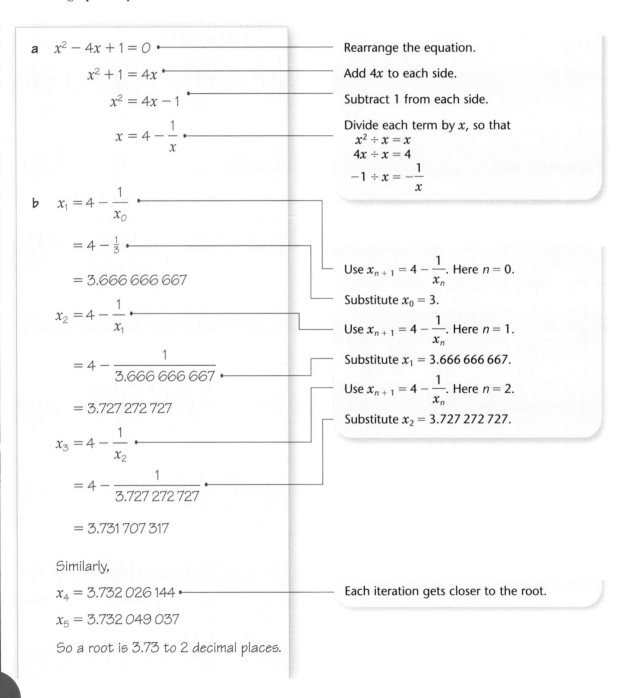

a $x^2 - 4x + 1 = 0$ — Rearrange the equation.

$x^2 + 1 = 4x$ — Add $4x$ to each side.

$x^2 = 4x - 1$ — Subtract 1 from each side.

$x = 4 - \dfrac{1}{x}$ — Divide each term by x, so that
$$x^2 \div x = x$$
$$4x \div x = 4$$
$$-1 \div x = -\dfrac{1}{x}$$

b $x_1 = 4 - \dfrac{1}{x_0}$

$= 4 - \dfrac{1}{3}$

$= 3.666\,666\,667$ — Use $x_{n+1} = 4 - \dfrac{1}{x_n}$. Here $n = 0$.

— Substitute $x_0 = 3$.

$x_2 = 4 - \dfrac{1}{x_1}$ — Use $x_{n+1} = 4 - \dfrac{1}{x_n}$. Here $n = 1$.

$= 4 - \dfrac{1}{3.666\,666\,667}$ — Substitute $x_1 = 3.666\,666\,667$.

$= 3.727\,272\,727$ — Use $x_{n+1} = 4 - \dfrac{1}{x_n}$. Here $n = 2$.

$x_3 = 4 - \dfrac{1}{x_2}$ — Substitute $x_2 = 3.727\,272\,727$.

$= 4 - \dfrac{1}{3.727\,272\,727}$

$= 3.731\,707\,317$

Similarly,

$x_4 = 3.732\,026\,144$ — Each iteration gets closer to the root.

$x_5 = 3.732\,049\,037$

So a root is 3.73 to 2 decimal places.

c

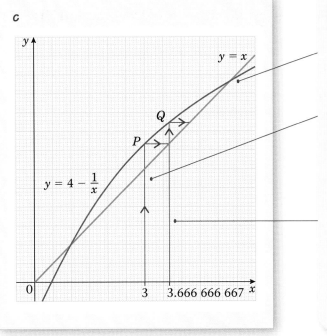

Draw the graphs of $y = x$ and $y = 4 - \dfrac{1}{x}$.

The graphs intersect when $x = 4 - \dfrac{1}{x}$.

Substitute $x = 3$ into $y = 4 - \dfrac{1}{x}$ so that $y = 3.666\,666\,667$. This is the same as moving vertically from $x = 3$ to P.

Let $x = 3.666\,666\,667$. This is the same as moving horizontally from P to the line $y = x$.

Substitute $x = 3.666\,666\,667$ into $y = 4 - \dfrac{1}{x}$ so that $y = 3.727\,272\,727$. This is the same as moving vertically from $x = 3.666\,666\,667$ to Q.

Let $x = 3.727\,272\,727$. This is the same as moving horizontally from Q to the line $y = x$.

Further iterations take you closer to the root.

The example above is a particular instance of this general result:

■ To solve an equation of the form f(x) = 0 by an iterative method, rearrange f(x) = 0 into a form x = g(x) and use the iterative formula x_{n+1} = g(x_n).

Example 6

a Show that $x^2 - 5x - 3 = 0$ can be written in the form

   i $x = \sqrt{5x + 3}$ **ii** $x = \dfrac{x^2 - 3}{5}$

b Show that the iteration formulae

   i $x_{n+1} = \sqrt{5x_n + 3}$ **ii** $x_{n+1} = \dfrac{x_n^2 - 3}{5}$

give different roots of the equation $x^2 - 5x - 3 = 0$. Start each iteration with $x_0 = 5$.

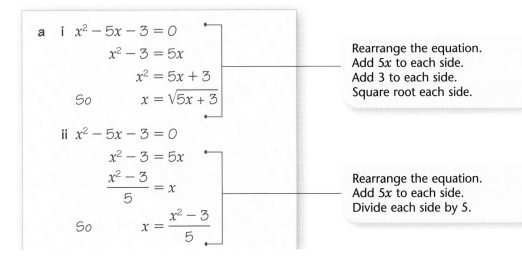

a   **i** $x^2 - 5x - 3 = 0$

          $x^2 - 3 = 5x$

          $x^2 = 5x + 3$

    So $x = \sqrt{5x + 3}$

Rearrange the equation.
Add $5x$ to each side.
Add 3 to each side.
Square root each side.

   ii $x^2 - 5x - 3 = 0$

          $x^2 - 3 = 5x$

          $\dfrac{x^2 - 3}{5} = x$

    So $x = \dfrac{x^2 - 3}{5}$

Rearrange the equation.
Add $5x$ to each side.
Divide each side by 5.

b i $x_1 = \sqrt{5x_0 + 3}$

$\quad = \sqrt{5(5) + 3}$

$\quad = 5.291\,502\,622$

$x_2 = \sqrt{5x_1 + 3}$

$\quad = \sqrt{5(5.291\,502\,622) + 3}$

$\quad = 5.427\,477\,601$

Similarly,

$x_3 = 5.489\,753$

$x_4 = 5.518\,039\,96$

$x_5 = 5.530\,840\,786$

$x_6 = 5.536\,623\,875$

So a root is 5.5 to 1 decimal place.

> Use $x_{n+1} = \sqrt{5x_n + 3}$. Here $n = 0$.
>
> Substitute $x_0 = 5$.
>
> Use $x_{n+1} = \sqrt{5x_n + 3}$. Here $n = 1$.
>
> Substitute $x_1 = 5.291\,502\,622$.
>
> Each iteration gets closer to a root, so the sequence $x_0, x_1, x_2, x_3, x_4 \ldots$ is **convergent**.

ii $x_1 = \dfrac{x_0^2 - 3}{5}$

$\quad = \dfrac{(5)^2 - 3}{5}$

$\quad = 4.4$

$x_2 = \dfrac{x_1^2 - 3}{5}$

$\quad = \dfrac{(4.4)^2 - 3}{5}$

$\quad = 3.272$

Similarly,

$x_3 = 1.541\,196\,8$

$x_4 = -0.124\,942\,484\,7$

$x_5 = -0.596\,877\,875\,1$

$x_6 = -0.528\,747\,360\,4$

$x_7 = -0.544\,085\,245\,8$

$x_8 = -0.540\,794\,249\,1$

So a root is -0.5 to 1 decimal place.

Each iteration formula gives a sequence that converges to a different root of the equation.

> Use $x_{n+1} = \dfrac{x_n^2 - 3}{5}$. Here $n = 0$.
>
> Substitute $x_0 = 5$.
>
> Use $x_{n+1} = \dfrac{x_n^2 - 3}{5}$. Here $n = 1$.
>
> Substitute $x_1 = 4.4$.
>
> The sequence $x_0, x_1, x_2, x_3, x_4 \ldots$ converges to a root.

■ Example 6 shows that different rearrangements of the equation $f(x) = 0$ give iteration formulae that **may** lead to different roots of the equation.

Example 7

a Show that $x^3 - 3x^2 - 2x + 5 = 0$ has a root in the interval $3 < x < 4$.

b Use the iteration formula $x_{n+1} = \sqrt{\dfrac{x_n^3 - 2x_n + 5}{3}}$ to find an approximation for the root of the

equation $x^3 - 3x^2 - 2x + 5 = 0$. Start with **i** $x_0 = 3$ **ii** $x_0 = 4$.

a Let $f(x) \equiv x^3 - 3x^2 - 2x + 5$

$f(3) = (3)^3 - 3(3)^2 - 2(3) + 5$

$\quad = 27 - 27 - 6 + 5$

$\quad = -1$

$f(4) = (4)^3 - 3(4)^2 - 2(4) + 5$

$\quad = 64 - 48 - 8 + 5$

$\quad = 13$

$f(3) < 0$ and $f(4) > 0$, so there is a change of sign

There is a root in the interval $3 < x < 4$.

> Show that $x^3 - 3x^2 - 2x + 5 = 0$ has a root between $x = 3$ and $x = 4$.
>
> Substitute $x = 3$ and $x = 4$ into the function.
>
> The graph crosses the x-axis between $x = 3$ and $x = 4$.

b i $x_1 = \sqrt{\dfrac{x_0^3 - 2x_0 - + 5}{3}}$

$\quad = \sqrt{\dfrac{(3)^3 - 2(3) + 5}{3}}$

$\quad = 2.943\,920\,289$

$x_2 = \sqrt{\dfrac{x_1^3 - 2x_1 + 5}{3}}$

$\quad = \sqrt{\dfrac{(2.943\ldots)^3 - (2.94\ldots) + 5}{3}}$

$\quad = 2.865\,084\,947$

Similarly,

$x_3 = 2.756\,113\,603$

$x_4 = 2.609\,192\,643$ etc.

$x_{14} = 1.203\,042\,309$

$x_{16} = 1.202\,094\,215$

So a root is 1.2 to 1 decimal place.

> Use $x_{n+1} = \sqrt{\dfrac{x_n^3 - 2x_n + 5}{3}}$. Here $n = 0$.
>
> Substitute $x_0 = 3$.
>
> Use $x_{n+1} = \sqrt{\dfrac{x_n^3 - 2x_n + 5}{3}}$. Here $n = 1$.
>
> Substitute $x_1 = 2.943\,920\,289$.
>
> The sequence $x_0, x_1, x_2, x_3, x_4 \ldots$ converges slowly to a root.
>
> This root is not in the interval $3 < x < 4$.

ii $x_1 = \sqrt{\dfrac{x_0^3 - 2x_0 + 5}{3}}$

$\quad = \sqrt{\dfrac{(4)^3 - 2(4) + 5}{3}}$

$\quad = 4.509\,249\,753$

$x_2 = \sqrt{\dfrac{x_1^3 - 2x_1 + 5}{3}}$

> Use $x_{n+1} = \sqrt{\dfrac{x_n^3 - 2x_n + 5}{3}}$. Here $n = 0$.
>
> Substitute $x_0 = 4$.
>
> Use $x_{n+1} = \sqrt{\dfrac{x_n^3 - 2x_n + 5}{3}}$. Here $n = 1$.

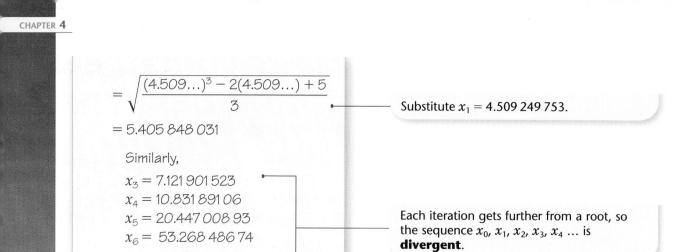

$$= \sqrt{\frac{(4.509\ldots)^3 - 2(4.509\ldots) + 5}{3}}$$

Substitute $x_1 = 4.509\ 249\ 753$.

$$= 5.405\ 848\ 031$$

Similarly,

$x_3 = 7.121\ 901\ 523$
$x_4 = 10.831\ 891\ 06$
$x_5 = 20.447\ 008\ 93$
$x_6 = 53.268\ 486\ 74$

Each iteration gets further from a root, so the sequence $x_0, x_1, x_2, x_3, x_4 \ldots$ is **divergent**.

No root is found.

■ Example 7 shows that even if you choose a value $x_0 = a$ that is close to a root, the sequence $x_0, x_1, x_2, x_3, x_4 \ldots$ does not necessarily converge to that root. In fact it might not converge to a root at all.

Exercise 4B

1 Show that $x^2 - 6x + 2 = 0$ can be written in the form:

a $x = \dfrac{x^2 + 2}{6}$ **b** $x = \sqrt{6x - 2}$ **c** $x = 6 - \dfrac{2}{x}$

2 Show that $x^3 + 5x^2 - 2 = 0$ can be written in the form:

a $x = \sqrt[3]{2 - 5x^2}$ **b** $x = \dfrac{2}{x^2} - 5$ **c** $x = \sqrt{\dfrac{2 - x^3}{5}}$

3 Rearrange $x^3 - 3x + 4 = 0$ into the form $x = \dfrac{x^3}{3} + a$, where the value of a is to be found.

4 Rearrange $x^4 - 3x^3 - 6 = 0$ into the form $x = \sqrt[3]{px^4 - 2}$, where the value of p is to be found.

5 **a** Show that the equation $x^3 - x^2 + 7 = 0$ can be written in the form $x = \sqrt[3]{x^2 - 7}$.

     b Use the iteration formula $x_{n+1} = \sqrt[3]{x_n^2 - 7}$, starting with $x_0 = 1$, to find x_2 to 1 decimal place.

6 **a** Show that the equation $x^3 + 3x^2 - 5 = 0$ can be written in the form $x = \sqrt{\dfrac{5}{x + 3}}$.

     b Use the iteration formula $x_{n+1} = \sqrt{\dfrac{5}{x_n + 3}}$, starting with $x_0 = 1$, to find x_4 to 3 decimal places.

7 **a** Show that the equation $x^6 - 5x + 3 = 0$ has a root between $x = 1$ and $x = 1.5$.

     b Use the iteration formula $x_{n+1} = \sqrt[5]{5 - \dfrac{3}{x_n}}$ to find an approximation for the root of the equation $x^6 - 5x + 3 = 0$, giving your answer to 2 decimal places.

8 **a** Rearrange the equation $x^2 - 6x + 1 = 0$ into the form $x = p - \dfrac{1}{x}$, where p is a constant to be found.

   b Starting with $x_0 = 3$, use the iteration formula $x_{n+1} = p - \dfrac{1}{x_n}$ with your value of p, to find x_3 to 2 decimal places.

9 **a** Show that the equation $x^3 - x^2 + 8 = 0$ has a root in the interval $(-2, -1)$.

   b Use a suitable iteration formula to find an approximation to 2 decimal places for the negative root of the equation $x^3 - x^2 + 8 = 0$.

10 **a** Show that $x^7 - 5x^2 - 20 = 0$ has a root in the interval $(1.6, 1.7)$.

   b Use a suitable iteration formula to find an approximation to 3 decimal places for the root of $x^7 - 5x^2 - 20 = 0$ in the interval $(1.6, 1.7)$.

Mixed exercise 4C

1 **a** Rearrange the cubic equation $x^3 - 6x - 2 = 0$ into the form $x = \pm\sqrt{a + \dfrac{b}{x}}$. State the values of the constants a and b.

   b Use the iterative formula $x_{n+1} = \sqrt{a + \dfrac{b}{x_n}}$ with $x_0 = 2$ and your values of a and b to find the approximate positive solution x_4 of the equation, to an appropriate degree of accuracy. Show all your intermediate answers. **E**

2 **a** By sketching the curves with equations $y = 4 - x^2$ and $y = e^x$, show that the equation $x^2 + e^x - 4 = 0$ has one negative root and one positive root.

   b Use the iteration formula $x_{n+1} = -(4 - e^{x_n})^{\frac{1}{2}}$ with $x_0 = -2$ to find in turn x_1, x_2, x_3 and x_4 and hence write down an approximation to the negative root of the equation, giving your answer to 4 decimal places.

   An attempt to evaluate the positive root of the equation is made using the iteration formula $x_{n+1} = (4 - e^{x_n})^{\frac{1}{2}}$ with $x_0 = 1.3$.

   c Describe the result of such an attempt. **E**

3 **a** Show that the equation $x^5 - 5x - 6 = 0$ has a root in the interval $(1, 2)$.

   b Stating the values of the constants p, q and r, use an iteration of the form $x_{n+1} = (px_n + q)^{\frac{1}{r}}$ an appropriate number of times to calculate this root of the equation $x^5 - 5x - 6 = 0$ correct to 3 decimal places. Show sufficient working to justify your final answer. **E**

4 $f(x) \equiv 5x - 4\sin x - 2$, where x is in radians.

   a Evaluate, to 2 significant figures, $f(1.1)$ and $f(1.15)$.

   b State why the equation $f(x) = 0$ has a root in the interval $(1.1, 1.15)$.

   An iteration formula of the form $x_{n+1} = p\sin x_n + q$ is applied to find an approximation to the root of the equation $f(x) = 0$ in the interval $(1.1, 1.15)$.

   c Stating the values of p and q, use this iteration formula with $x_0 = 1.1$ to find x_4 to 3 decimal places. Show the intermediate results in your working. **E**

5 $f(x) \equiv 2\sec x + 2x - 3$, where x is in radians.

   a Evaluate $f(0.4)$ and $f(0.5)$ and deduce the equation $f(x) = 0$ has a solution in the interval $0.4 < x < 0.5$.

   b Show that the equation $f(x) = 0$ can be arranged in the form $x = p + \dfrac{q}{\cos x}$, where p and q are constants, and state the value of p and the value of q.

   c Using the iteration formula $x_{n+1} = p + \dfrac{q}{\cos x_n}$, $x_0 = 0.4$, with the values of p and q found in part **b**, calculate x_1, x_2, x_3 and x_4, giving your final answer to 4 decimal places. **E**

6 $f(x) \equiv e^{0.8x} - \dfrac{1}{3 - 2x}$, $x \neq \frac{3}{2}$

   a Show that the equation $f(x) = 0$ can be written as $x = 1.5 - 0.5e^{-0.8x}$.

   b Use the iteration formula $x_{n+1} = 1.5 - 0.5e^{-0.8x_n}$ with $x_0 = 1.3$ to obtain x_1, x_2 and x_3. Give the value of x_3, an approximation to a root of $f(x) = 0$, to 3 decimal places.

   c Show that the equation $f(x) = 0$ can be written in the form $x = p \ln(3 - 2x)$, stating the value of p.

   d Use the iteration formula $x_{n+1} = p \ln(3 - 2x_n)$ with $x_0 = -2.6$ and the value of p found in part **c** to obtain x_1, x_2 and x_3. Give the value of x_3, an approximation to the second root of $f(x) = 0$, to 3 decimal places. **E**

7 **a** Use the iteration $x_{n+1} = (3x_n + 3)^{\frac{1}{3}}$ with $x_0 = 2$ to find, to 3 significant figures, x_4.

   The only real root of the equation $x^3 - 3x - 3 = 0$ is α. It is given that, to 3 significant figures, $\alpha = x_4$.

   b Use the substitution $y = 3^x$ to express $27^x - 3^{x+1} - 3 = 0$ as a cubic equation.

   c Hence, or otherwise, find an approximate solution to the equation $27^x - 3^{x+1} - 3 = 0$, giving your answer to 2 significant figures. **E**

8 The equation $x^x = 2$ has a solution near $x = 1.5$.

   a Use the iteration formula $x_{n+1} = 2^{\frac{1}{x_n}}$ with $x_0 = 1.5$ to find the approximate solution x_5 of the equation. Show the intermediate iterations and give your final answer to 4 decimal places.

   b Use the iteration formula $x_{n+1} = 2x_n^{(1 - x_n)}$ with $x_0 = 1.5$ to find x_1, x_2, x_3, x_4. Comment briefly on this sequence. **E**

9 **a** Show that the equation $2^{1-x} = 4x + 1$ can be arranged in the form $x = \frac{1}{2}(2^{-x}) + q$, stating the value of the constant q.

   b Using the iteration formula $x_{n+1} = \frac{1}{2}(2^{-x_n}) + q$ with $x_0 = 0.2$ and the value of q found in part **a**, find x_1, x_2, x_3 and x_4. Give the value of x_4, to 4 decimal places. **E**

10 The curve with equation $y = \ln(3x)$ crosses the x-axis at the point P $(p, 0)$.

   a Sketch the graph of $y = \ln(3x)$, showing the exact value of p.

   The normal to the curve at the point Q, with x-coordinate q, passes through the origin.

   b Show that $x = q$ is a solution of the equation $x^2 + \ln 3x = 0$.

   c Show that the equation in part **b** can be rearranged in the form $x = \frac{1}{3}e^{-x^2}$.

   d Use the iteration formula $x_{n+1} = \frac{1}{3}e^{-x_n^2}$, with $x_0 = \frac{1}{3}$, to find x_1, x_2, x_3 and x_4. Hence write down, to 3 decimal places, an approximation for q. **E**

11 a Copy this sketch of the curve with equation $y = e^{-x} - 1$.
On the same axes sketch the graph of $y = \frac{1}{2}(x - 1)$, for $x \geq 1$,
and $y = -\frac{1}{2}(x - 1)$, for $x < 1$. Show the coordinates of the
points where the graph meets the axes.

The x-coordinate of the point of intersection of the graphs
is α.

b Show that $x = \alpha$ is a root of the equation $x + 2e^{-x} - 3 = 0$.

c Show that $-1 < \alpha < 0$.

The iterative formula $x_{n+1} = -\ln[\frac{1}{2}(3 - x_n)]$ is used to solve the equation $x + 2e^{-x} - 3 = 0$.

d Starting with $x_0 = -1$, find the values of x_1 and x_2.

e Show that, to 2 decimal places, $\alpha = -0.58$.

E

Summary of key points

1 If you find an interval in which f(x) changes sign, and f(x) is continuous in that interval, then the interval must contain a root of the equation f(x) = 0.

2 To solve an equation of the form f(x) = 0 by an iterative method, rearrange f(x) = 0 into a form $x = $ g(x) and use the iterative formula $x_{n+1} = $ g(x_n).

3 Different rearrangements of the equation f(x) = 0 give iteration formulae that **may** lead to different roots of the equation.

4 If you choose a value $x_0 = a$ for the starting value in an iteration formula, and $x_0 = a$ is close to a root of the equation f(x) = 0, then the sequence $x_0, x_1, x_2, x_3, x_4 \ldots$ does not necessarily converge to that root. In fact it might not converge to a root at all.

Review Exercise

1 Simplify

  a $\dfrac{2x^2 - 7x - 15}{x^2 - 25}$

  b $\dfrac{x^3 + 1}{x + 1}$.

2 Express $\dfrac{4x}{x^2 - 2x - 3} + \dfrac{1}{x^2 + x}$ as a single fraction, giving your answer in its simplest form.

3 Express $\dfrac{2x^2 + 3x}{(2x + 3)(x - 2)} - \dfrac{6}{x^2 - x - 2}$ as a single fraction in its simplest form. **E**

4 **a** Given that

    $16x^3 - 36x^2 - 12x + 5$
    $\equiv (2x + 1)(8x^2 + ax + b),$

    find the value of a and the value of b.

  b Hence, or otherwise, simplify

    $\dfrac{16x^3 - 36x^2 - 12x + 5}{4x - 1}$

5 $f(x) = 1 - \dfrac{3}{x + 2} + \dfrac{3}{(x + 2)^2}, \quad x \neq -2$

  a Show that $f(x) = \dfrac{x^2 + x + 1}{(x + 2)^2}, \quad x \neq -2$

  b Show that $x^2 + x + 1 > 0$ for all values of x.

  c Show that $f(x) > 0$ for all values of x, $x \neq -2$. **E**

6 **a** Show that

    $\dfrac{4}{(x + 1)^2} - \dfrac{1}{(x + 1)} - \dfrac{1}{2} = \dfrac{5 - 4x - x^2}{2(x + 1)^2}.$

  b Hence solve

    $\dfrac{4}{(x + 1)^2} < \dfrac{1}{(x + 1)} + \dfrac{1}{2}, \ x \neq -1.$

7 $f(x) = \dfrac{x}{x + 3} - \dfrac{x + 24}{2x^2 + 5x - 3},$

    $\{x \in \mathbb{R}, x > \tfrac{1}{2}\}.$

  a Show that $f(x) = \dfrac{2(x - 4)}{2x - 1},$

    $\{x \in \mathbb{R}, x > \tfrac{1}{2}\}.$

  b Find $f^{-1}(x)$.

8 The graph of the increasing function f passes through the points A(0, −2), B(3, 0) and C(5, 2), as shown.

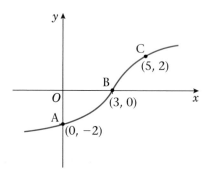

a Sketch the graph of f^{-1}, showing the images of A, B and C.

The function g is defined by

g: $x \rightarrow \sqrt{x^2 + 2}$, $x \in \mathbb{R}$

b Find **i** $fg\sqrt{23}$, **ii** $gf(0)$.

9 The functions f and g are defined by

f: $x \rightarrow 3x + 4$, $x \in \mathbb{R}$, $x > 0$,

g: $x \rightarrow \dfrac{x}{x - 2}$, $x \in \mathbb{R}$, $x > 2$.

a Find the inverse function $f^{-1}(x)$, stating its domain.

b Find the exact value of $gf(\frac{1}{2})$.

c State the range of g.

d Find $g^{-1}(x)$, stating its domain.

10 The function f is defined by

$f : x \rightarrow \dfrac{5x + 1}{x^2 + x - 2} - \dfrac{3}{x + 2}$, $x > 1$.

a Show that $f(x) = \dfrac{2}{x - 1}$, $x > 1$.

b Find $f^{-1}(x)$.

The function g is defined by

$g : x \rightarrow x^2 + 5$, $x \in \mathbb{R}$

c Solve $fg(x) = \frac{1}{4}$. **E**

11 The functions f and g are defined by

f: $x \rightarrow (x - 4)^2 - 16$, $x \in \mathbb{R}$, $x > 0$,

g: $x \rightarrow \dfrac{8}{1 - x}$, $x \in \mathbb{R}$, $x < 1$.

a Find the range of f.

b Explain why, with the given domain for f, $f^{-1}(x)$ does not exist.

c Show that $fg(x) = \dfrac{64x}{(1 - x)^2}$.

d Find $g^{-1}(x)$, stating its domain.

12 The function $f(x)$ is defined by

$f(x) = \begin{cases} -2(x + 1) & -2 \leqslant x \leqslant -1 \\ (x + 1)(2 - x) & -1 < x \leqslant 2 \end{cases}$

a Sketch the graph of $f(x)$.

b Write down the range of f.

c Find the values of x for which $f(x) < 2$.

13 **a** Express $4x^2 - 4x - 3$ in the form $(ax - b)^2 - c$, where a, b and c are positive constants to be found.

The function f is defined by

f: $x \rightarrow 4x^2 - 4x - 3$, $\{x \in \mathbb{R},\ x \geqslant \frac{1}{2}\}$.

b Sketch the graph of f.

c Sketch the graph of f^{-1}.

d Find $f^{-1}(x)$, stating its domain.

14 The functions f and g are defined by

f: $x \rightarrow \dfrac{x + 2}{x}$, $x \in \mathbb{R}$, $x \neq 0$.

g: $x \rightarrow \ln(2x - 5)$, $x \in \mathbb{R}$, $x > 2\frac{1}{2}$.

a Sketch the graph of f.

b Show that $f^2(x) = \dfrac{3x + 2}{x + 2}$.

[$f^2(x)$ means $ff(x)$]

c Find the exact value of $gf\left(\frac{1}{4}\right)$.

d Find $g^{-1}(x)$, stating its domain.

15 Solve the following equations, giving your answers to 3 significant figures.

a $3e^{(2x+5)} = 4$

b $3^x = 5^{1-x}$

c $2\ln(2x - 1) = 1 + \ln 7$

16 Find the exact solutions to the equations

a $\ln x + \ln 3 = \ln 6$

b $e^x + 3e^{-x} = 4$ **E**

17 The function f is defined by

f: $x \rightarrow 3 - \ln(x + 2)$, $x \in \mathbb{R}$, $x > -2$.

The graph of $y = f(x)$ crosses the x-axis at the point A and crosses the y-axis at the point B.

a Find the exact coordinates of A and B.

b Sketch the graph of $y = f(x)$, $x > -2$.

18 The graph of the function

$f(x) = 144 - 36e^{-2x}$, $x \in \mathbb{R}$

has an asymptote $y = k$, and crosses the x and y axes at A and B respectively, as shown overleaf.

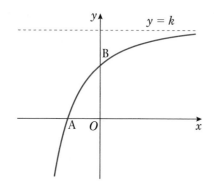

a Write down the value of k and the y-coordinate of B.

b Express the x-coordinate of A in terms of ln2.

19 [Part **d** requires the differentiation of e^{ax}, see Ch 8]

The functions f and g are defined by
$$f : x \rightarrow 2x + \ln 2, \quad x \in \mathbb{R}$$
$$g : x \rightarrow e^{2x}, \quad x \in \mathbb{R}.$$

a Prove that the composite function gf is
$$gf : x \rightarrow 4e^{4x}, \quad x \in \mathbb{R}.$$

b Sketch the curve with equation $y = gf(x)$, and show the coordinates of the point where the curve crosses the y-axis.

c Write down the range of gf.

d Find the value of x for which $\frac{d}{dx}[gf(x)] = 3$, giving your answer to 3 significant figures. **E**

20 **a** Show that $e^x - e^{-x} = 4$ can be rewritten in the form $e^{2x} - 4e^x - 1 = 0$

b Hence find the exact value of the real solution of $e^x - e^{-x} = 4$.

c For this value of x, find the exact value of $e^x + e^{-x}$.

21 At time $t = 0$, a lake is stocked with k fish. The number, n, of fish in the lake at time t days can be represented by the equation
$$n = 3000 + 1450e^{0.04t}.$$

a State the value of k.

b Calculate the increase in the population of fish 3 weeks after stocking the lake.

c Find how many days pass, from the day the lake was stocked, before the number of fish increases to over 7000.

22 A heated metal ball S is dropped into a liquid. As S cools its temperature, $T\,°C$, t minutes after it enters the liquid is given by
$$T = 400e^{-0.05t} + 25, \quad t \geqslant 0.$$

a Find the temperature of S as it enters the liquid.

b Find how long S is in the liquid before its temperature drops to 300 °C. Give your answer to 3 significant figures.

c Find the rate, in °C per minute to 3 significant figures, at which the temperature of S is decreasing at the instant $t = 50$.

d With reference to the equation given above, explain why the temperature of S can never drop to 20 °C **E**

23 A breeding programme for a particular animal is being monitored. Initially there were k breeding pairs in the survey. A suggested model for the number of breeding pairs, n, after t years is
$$n = \frac{400}{1 + 9e^{-\frac{1}{9}t}}.$$

a Find the value of k.

b Show that the above equation can be written in the form $t = 9 \ln \left(\dfrac{9n}{400 - n} \right)$

c Hence, or otherwise, calculate the number of years, according to the model, after which the number of breeding pairs will first exceed 100.

The model predicts that the number of breeding pairs cannot exceed the value A.

d Find the value of A.

24 $f(x) = x^3 - \dfrac{1}{x} - 2, \quad x \neq 0.$

a Show that the equation $f(x) = 0$ has a root between 1 and 2.

An approximation for this root is found using the iteration formula

$$x_{n+1} = \left(2 + \frac{1}{x_n}\right)^{\frac{1}{3}}, \text{ with } x_0 = 1.5.$$

b By calculating the values of x_1, x_2, x_3 and x_4 find an approximation to this root, giving your answer to 3 decimal places.

c By considering the change of sign of f(x) in a suitable interval, verify that your answer to part **b** is correct to 3 decimal places. **E**

25 a By sketching the graphs of $y = -x$ and $y = \ln x$, $x > 0$, on the same axes, show that the solution to the equation $x + \ln x = 0$ lies between 0 and 1.

b Show that $x + \ln x = 0$ may be written in the form $x = \dfrac{(2x - \ln x)}{3}$.

c Use the iterative formula

$$x_{n+1} = \frac{(2x_n - \ln x_n)}{3}, \quad x_0 = 1,$$

to find the solution of $x + \ln x = 0$ correct to 5 decimal places.

26 Show that the equation $e^{2x} - 8x = 0$ has a root k between $x = 1$ and $x = 2$.

The iterative formula

$$x_n = \tfrac{1}{2}\ln 8x, \quad x_0 = 1.2,$$

is used to find to find an approximation for k.

a Calculate the values of x_1, x_2 and x_3, giving your answers to 3 decimal places.

b Show that, to 3 decimal places, $k = 1.077$.

c Deduce the value, to 2 decimal places, of one of the roots of $e^x = 4x$.

27 The curve C has equation $y = x^5 - 1$. The tangent to C at the point $P(-1, -2)$ meets the curve again at the point Q, whose x-coordinate is k.

a Show that k is a root of the equation $x^5 - 5x - 4 = 0$.

b Show that $x^5 - 5x - 4 = 0$ can be rearranged in the form $x = \sqrt[4]{5 + \dfrac{4}{x}}$.

The iterative formula

$$x_{n+1} = \sqrt[4]{5 + \frac{4}{x_n}}, \quad x_0 = 1.5,$$

is used to find to find an approximation for k.

c Write down the values of x_1, x_2, x_3 and x_4, giving your answers to 5 significant figures.

d Show that $k = 1.6506$ correct to 5 significant figures.

28 f(x) = $2x^3 - x - 4$.

a Show that the equation f(x) = 0 can be written as

$$x = \sqrt{\left(\frac{2}{x} + \frac{1}{2}\right)}.$$

The equation $2x^3 - x - 4 = 0$ has a root between 1.35 and 1.4.

b Use the iteration formula

$$x_{n+1} = \sqrt{\left(\frac{2}{x_n} + \frac{1}{2}\right)},$$

with $x_0 = 1.35$, to find, to 2 decimal places, the value of x_1, x_2 and x_3.

The only real root of f(x) = 0 is α.

c By choosing a suitable interval, prove that $\alpha = 1.392$, to 3 decimal places. **E**

29 The function f is defined by
f : $x \rightarrow -5 + 4e^{2x}$, $x \in \mathbb{R}$, $x > 0$.

a Show that the inverse function of f is defined by

$$f^{-1}: x \rightarrow \frac{1}{2}\ln\left(\frac{x+5}{4}\right),$$

and write down the domain of f^{-1}.

b Write down the range of f^{-1}.

The graphs of $y = \tfrac{1}{2}x$ and $y = f^{-1}(x)$, drawn on the same axes, meet at the point with the x-coordinate k.

The iterative formula

$$x_{n+1} = \ln\left(\frac{x_n + 5}{4}\right), \quad x_0 = 0.3,$$

is used to find to find an approximation for k.

c Calculate the values of x_1 and x_2, giving your answers to 4 decimal places.

d Continue the iterative process until there are two values which are the same to 4 decimal places.

e Prove that this value does give k, correct to 4 decimal places.

30 The graph of the function f, defined by
$$\text{f: } x \rightarrow \frac{1}{1 + x^2}, \quad x \in \mathbb{R}, \quad x \geq 0,$$
is shown.

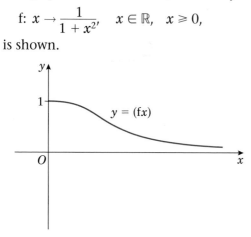

a Copy the sketch and add to it the graph of $y = f^{-1}(x)$, showing the coordinates of the point where it meets the x-axis.

The two curves meet in the point A, with x-coordinate k.

b Explain why k is a solution of the equation $x = \dfrac{1}{1 + x^2}$.

The iterative formula
$$x_{n+1} = \frac{1}{1 + x_n^2}, \quad x_0 = 0.7$$
is used to find an approximation for k.

c Calculate the values of x_1, x_2, x_3 and x_4, giving your answers to 4 decimal places.

d Show that $k = 0.682$, correct to 3 decimal places.

After completing this chapter you should be able to

1 sketch the graph of the modulus function $y = |f(x)|$

2 sketch the graph of the function $y = f(|x|)$

3 solve equations involving the modulus function

4 apply a combination of two (or more) transformations to the same curve

5 sketch transformations of the graph $y = f(x)$.

Transforming graphs of functions

An example of the modulus graph can be found in electricity generation. Electricty is generated as alternating current. Its shape is that of a sine curve. In some appliances, such as mobile phone chargers, it goes through a series of changes. One of these changes uses a rectifier to transform the $y = \sin x$ graph into $y = |\sin x|$. A capacitor then 'smoothes' out the wave to convert it into the direct current that is used in some appliances.

Wind farms provide a renewable source of electricity.

5.1 You need to be able to sketch the graph of the modulus function $y = |f(x)|$.

■ The modulus of a number a, written as $|a|$, is its **positive** numerical value.

So, for example, $|5| = 5$ and also $|-5| = 5$.

It is sometimes known as the **absolute value**, and is shown on the display of some calculators as, for example, 'Abs -5' or 'Abs(-5)'. If your calculator has a modulus or absolute value button, make sure you understand how to use it.

■ A modulus function is, in general, a function of the type $y = |f(x)|$.
  When $f(x) \geqslant 0$, $|f(x)| = f(x)$.
  When $f(x) < 0$, $|f(x)| = -f(x)$.

Example 1

Sketch the graph of $y = |x|$.

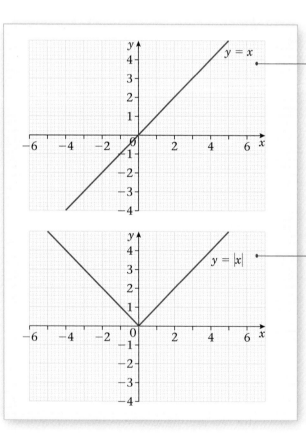

Step 1
Sketch the graph of $y = x$.
(Ignore the modulus.)

Step 2
For the part of the line below the x-axis (the negative values of y), reflect in the x-axis. For example this will change the y-value -3 into the y-value 3.

Important
If you do steps 1 and 2 above on the same diagram, make sure that you clearly show that you have deleted the part of the graph below the x-axis.

Example 2

Sketch the graph of $y = |3x - 2|$.

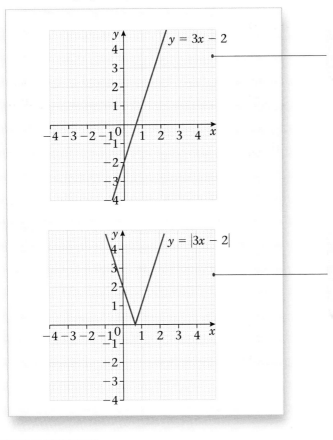

Step 1
Sketch the graph of $y = 3x - 2$.
(Ignore the modulus.)

Step 2
For the part of the line below the x-axis (the negative values of y), reflect in the x-axis. For example, this will change the y-value -2 into the y-value 2.

Example 3

Sketch the graph of $y = |x^2 - 3x - 10|$.

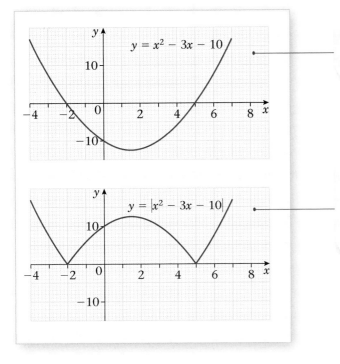

Step 1
Sketch the graph of $y = x^2 - 3x - 10$.
(Ignore the modulus.)

Step 2
For the part of the curve below the x-axis (the negative values of y), reflect in the x-axis. For example, this will change the y-value -3 into the y-value 3.

Example 4

The diagram on the right shows the graph of $y = f(x)$.
Sketch the graph of $y = |f(x)|$.

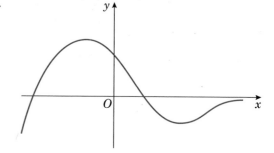

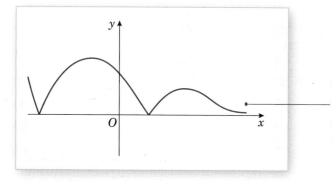

As in the previous examples, the part of the curve below the x-axis must be reflected in the x-axis. The graph of $y = |f(x)|$ looks like this.

Example 5

Sketch the graph of $y = |\sin x|$, $0 \leqslant x \leqslant 2\pi$.

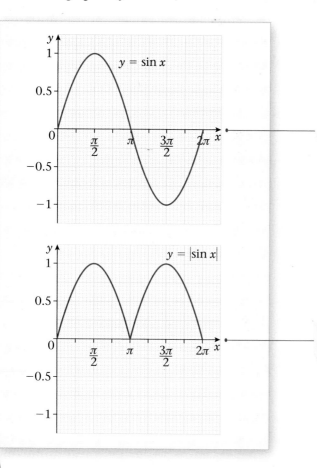

First draw the graph of $y = \sin x$.

As before, reflect the part of the curve below the x-axis in the x-axis.

Exercise 5A

1 Sketch the graph of each of the following. In each case, write down the coordinates of any points at which the graph meets the coordinate axes.

a $y = |x - 1|$

b $y = |2x + 3|$

c $y = |\frac{1}{2}x - 5|$

d $y = |7 - x|$

e $y = |x^2 - 7x - 8|$

f $y = |x^2 - 9|$

g $y = |x^3 + 1|$

h $y = \left| \dfrac{12}{x} \right|$

i $y = -|x|$

j $y = -|3x - 1|$

2 Sketch the graph of each of the following. In each case, write down the coordinates of any points at which the graph meets the coordinate axes.

a $y = |\cos x|,\ 0 \leqslant x \leqslant 2\pi$

b $y = |\ln x|,\ x > 0$

c $y = |2^x - 2|$

d $y = |100 - 10^x|$

e $y = |\tan 2x|,\ 0 < x < 2\pi$

5.2 You need to be able to sketch the graph of the function $y = f(|x|)$.

For the function $y = f(|x|)$, the value of y at, for example, $x = -5$ is the same as the value of y at $x = 5$. This is because $f(|-5|) = f(5)$.

Example 6

Sketch the graph of $y = |x| - 2$.

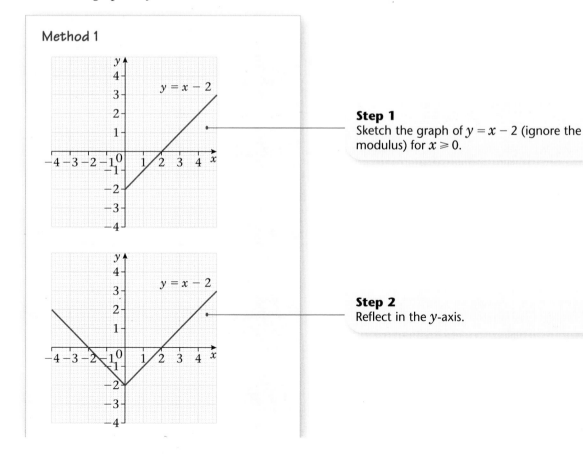

Method 1

$y = x - 2$

Step 1
Sketch the graph of $y = x - 2$ (ignore the modulus) for $x \geqslant 0$.

$y = x - 2$

Step 2
Reflect in the y-axis.

Method 2

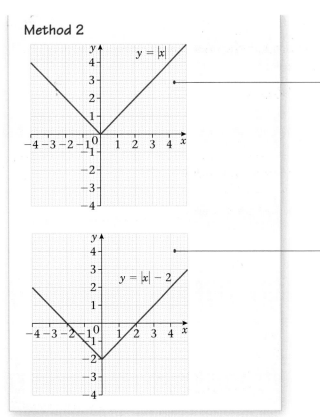

Step 1
Sketch the graph of $y = |x|$.

Step 2
Vertical translation of -2 units.
(See transformations of curves in Book C1,
Chapter 4.)

Example 7

Sketch the graph of $y = 4|x| - |x|^3$.

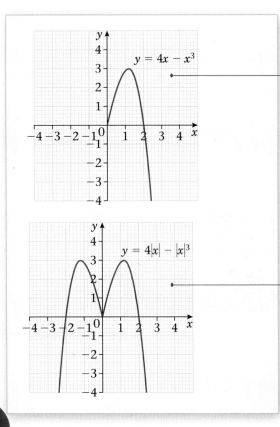

Step 1
Sketch the graph of $y = 4x - x^3$ (ignore the
modulus) for $x \geq 0$.

Step 2
Reflect in the y-axis.

Exercise 5B

Sketch the graph of each of the following. In each case, write down the coordinates of any points at which the graph meets the coordinate axes.

1 $y = 2|x| + 1$

2 $y = |x|^2 - 3|x| - 4$

3 $y = \sin|x|, \; -2\pi \leqslant x \leqslant 2\pi$

4 $y = 2^{|x|}$

5.3 You need to be able to solve equations involving a modulus.

Solutions can come from either the 'original' or the 'reflected' part of the graph.

Example 8

Solve the equation $|2x - \frac{3}{2}| = 3$.

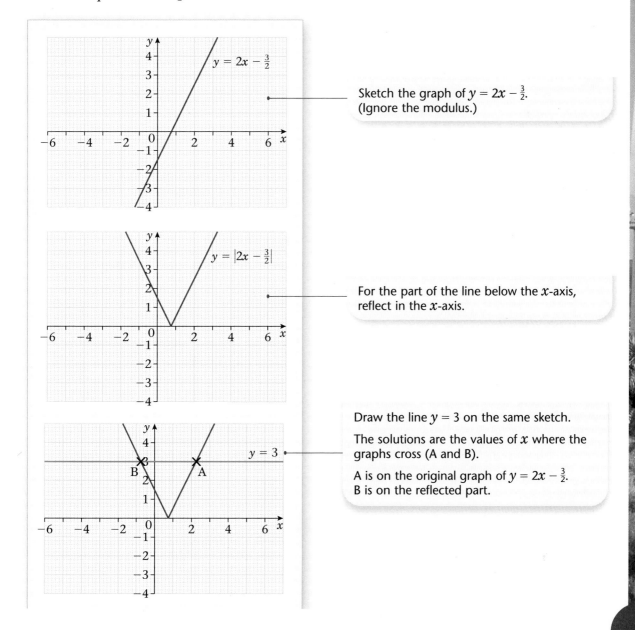

Sketch the graph of $y = 2x - \frac{3}{2}$.
(Ignore the modulus.)

For the part of the line below the x-axis, reflect in the x-axis.

Draw the line $y = 3$ on the same sketch.

The solutions are the values of x where the graphs cross (A and B).

A is on the original graph of $y = 2x - \frac{3}{2}$.
B is on the reflected part.

At A, $2x - \frac{3}{2} = 3$ ● ————— Original: Use $2x - \frac{3}{2}$.

$$2x = \frac{9}{2}$$

$$x = \frac{9}{4} = 2\frac{1}{4}$$

At B, $-(2x - \frac{3}{2}) = 3$ ● ————— When $f(x) < 0$, $|f(x)| = -f(x)$, so, as it is reflected, use $-(2x - \frac{3}{2})$.

$$-2x + \frac{3}{2} = 3$$

$$-2x = \frac{3}{2}$$

$$x = -\frac{3}{4}$$

The solutions to the equation are $x = -\frac{3}{4}$ and $x = \frac{9}{4}$.

Example 9

a On the same diagram, sketch the graphs of $y = |5x - 2|$ and $y = |2x|$.

b Solve the equation $|5x - 2| = |2x|$.

a

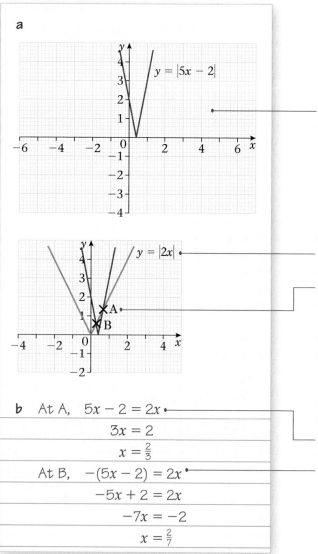

Sketch the graph of $y = |5x - 2|$. (As usual, for the part of $y = 5x - 2$ that is below the x-axis, reflect in the x-axis.)

On the same diagram, sketch the graph of $y = |2x|$.

The solutions for part **b** are the values of x where the 2 graphs intersect.

Intersection point A is on the original graph of $y = 5x + 2$, and on the original graph of $y = 2x$.

Intersection point B is on the reflected part of $y = 5x - 2$, and on the original graph of $y = 2x$.

b At A, $5x - 2 = 2x$ ● ————— Original: Use $5x - 2$ and $2x$.

$$3x = 2$$

$$x = \frac{2}{3}$$

At B, $-(5x - 2) = 2x$ ● ————— Reflected: Use $-(5x - 2)$

$$-5x + 2 = 2x$$ Original: Use $2x$.

$$-7x = -2$$

$$x = \frac{2}{7}$$

Example 10

a On the same diagram, sketch the graphs of $y = |x^2 - 2x|$ and $y = \frac{1}{4} - 2x$.

b Solve the equation $|x^2 - 2x| = \frac{1}{4} - 2x$.

a

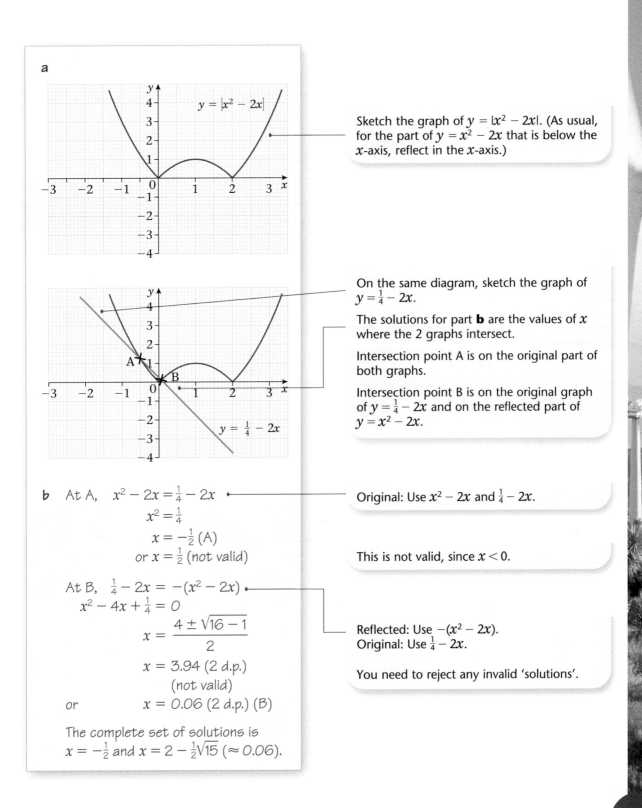

Sketch the graph of $y = |x^2 - 2x|$. (As usual, for the part of $y = x^2 - 2x$ that is below the x-axis, reflect in the x-axis.)

On the same diagram, sketch the graph of $y = \frac{1}{4} - 2x$.

The solutions for part **b** are the values of x where the 2 graphs intersect.

Intersection point A is on the original part of both graphs.

Intersection point B is on the original graph of $y = \frac{1}{4} - 2x$ and on the reflected part of $y = x^2 - 2x$.

b At A, $x^2 - 2x = \frac{1}{4} - 2x$

$x^2 = \frac{1}{4}$

$x = -\frac{1}{2}$ (A)

or $x = \frac{1}{2}$ (not valid)

Original: Use $x^2 - 2x$ and $\frac{1}{4} - 2x$.

This is not valid, since $x < 0$.

At B, $\frac{1}{4} - 2x = -(x^2 - 2x)$

$x^2 - 4x + \frac{1}{4} = 0$

$x = \dfrac{4 \pm \sqrt{16 - 1}}{2}$

$x = 3.94$ (2 d.p.)
(not valid)

or $x = 0.06$ (2 d.p.) (B)

Reflected: Use $-(x^2 - 2x)$.
Original: Use $\frac{1}{4} - 2x$.

You need to reject any invalid 'solutions'.

The complete set of solutions is
$x = -\frac{1}{2}$ and $x = 2 - \frac{1}{2}\sqrt{15}$ (≈ 0.06).

Exercise 5C

1 On the same diagram, sketch the graphs of $y = -2x$ and $y = |\frac{1}{2}x - 2|$. Solve the equation $-2x = |\frac{1}{2}x - 2|$.

2 On the same diagram, sketch the graphs of $y = |x|$ and $y = |-4x - 5|$. Solve the equation $|x| = |-4x - 5|$.

3 On the same diagram, sketch the graphs of $y = 3x$ and $y = |x^2 - 4|$. Solve the equation $3x = |x^2 - 4|$.

4 On the same diagram, sketch the graphs of $y = |x| - 1$ and $y = -|3x|$. Solve the equation $|x| - 1 = -|3x|$.

5 On the same diagram, sketch the graphs of $y = 24 + 2x - x^2$ and $y = |5x - 4|$. Solve the equation $24 + 2x - x^2 = |5x - 4|$. (Answers to 2 d.p. where appropriate).

5.4 You need to be able to apply a combination of two (or more) transformations to the same curve.

In Book C1, Chapter 4, you saw how to apply various transformations to curves. To summarise these:

- ① $f(x + a)$ is a horizontal translation of $-a$
  ② $f(x) + a$ is a vertical translation of $+a$
  ③ $f(ax)$ is a horizontal stretch of scale factor $\dfrac{1}{a}$
  ④ $af(x)$ is a vertical stretch of scale factor a

Example 11

Sketch the graph of $y = (x - 2)^2 + 3$.

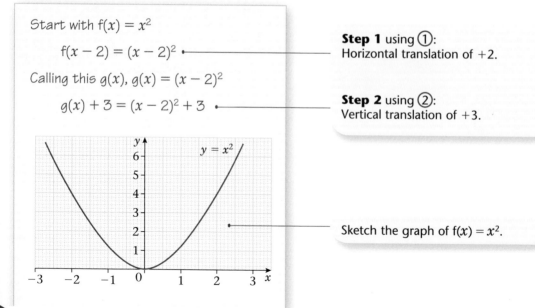

Start with $f(x) = x^2$

$$f(x - 2) = (x - 2)^2$$

Step 1 using ①:
Horizontal translation of $+2$.

Calling this $g(x)$, $g(x) = (x - 2)^2$

$$g(x) + 3 = (x - 2)^2 + 3$$

Step 2 using ②:
Vertical translation of $+3$.

Sketch the graph of $f(x) = x^2$.

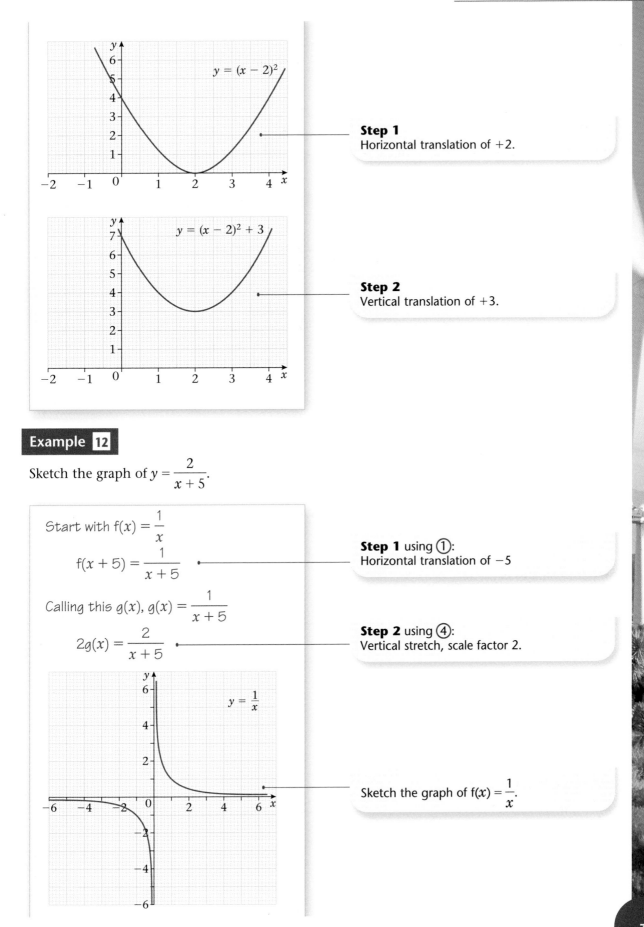

Step 1
Horizontal translation of $+2$.

Step 2
Vertical translation of $+3$.

Example 12

Sketch the graph of $y = \dfrac{2}{x + 5}$.

Start with $f(x) = \dfrac{1}{x}$

$f(x + 5) = \dfrac{1}{x + 5}$

Step 1 using ①:
Horizontal translation of -5

Calling this $g(x)$, $g(x) = \dfrac{1}{x + 5}$

$2g(x) = \dfrac{2}{x + 5}$

Step 2 using ④:
Vertical stretch, scale factor 2.

Sketch the graph of $f(x) = \dfrac{1}{x}$.

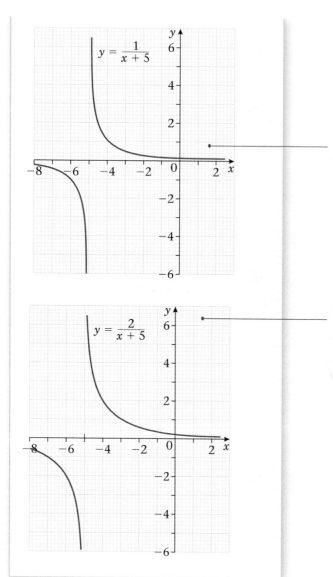

Step 1
Horizontal translation of -5.

Step 2
Vertical stretch, scale factor 2.

Notice what happens to a point such as $(-4, 1)$ …. It goes to $(-4, 2)$.

Example 13

Sketch the graph of $y = \cos 2x - 1$.

Start with $f(x) = \cos x$

$\qquad f(2x) = \cos 2x$

Calling this $g(x)$, $g(x) = \cos 2x$

$\qquad g(x) - 1 = \cos 2x - 1$

Step 1 using ③:
Horizontal stretch, scale factor $\frac{1}{2}$.

Step 2 using ②:
Vertical translation of -1.

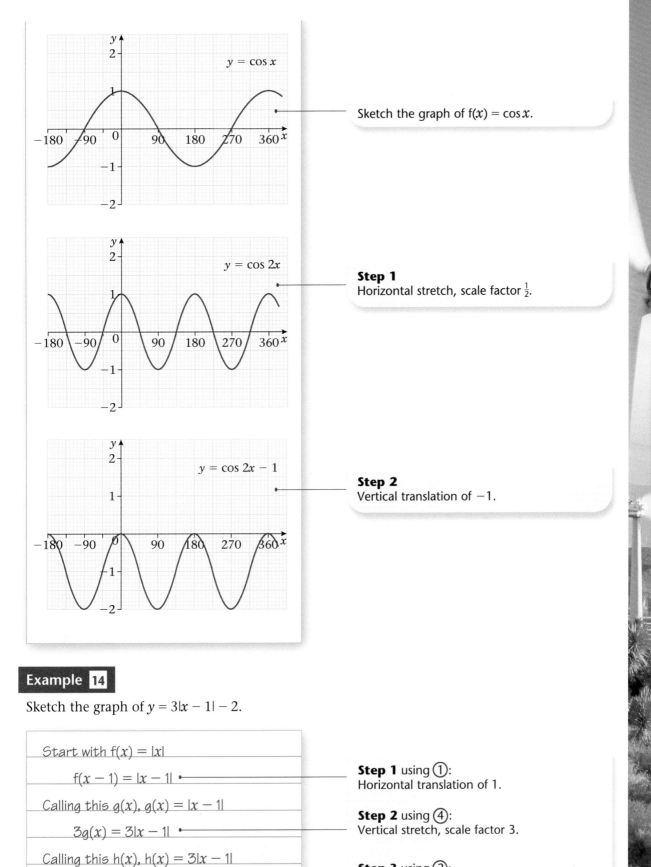

Sketch the graph of $f(x) = \cos x$.

Step 1
Horizontal stretch, scale factor $\frac{1}{2}$.

Step 2
Vertical translation of -1.

Example 14

Sketch the graph of $y = 3|x - 1| - 2$.

Start with $f(x) = |x|$

$f(x - 1) = |x - 1|$

Calling this $g(x)$, $g(x) = |x - 1|$

$3g(x) = 3|x - 1|$

Calling this $h(x)$, $h(x) = 3|x - 1|$

$h(x) - 2 = 3|x - 1| - 2$

Step 1 using ①:
Horizontal translation of 1.

Step 2 using ④:
Vertical stretch, scale factor 3.

Step 3 using ②:
Vertical translation of -2.

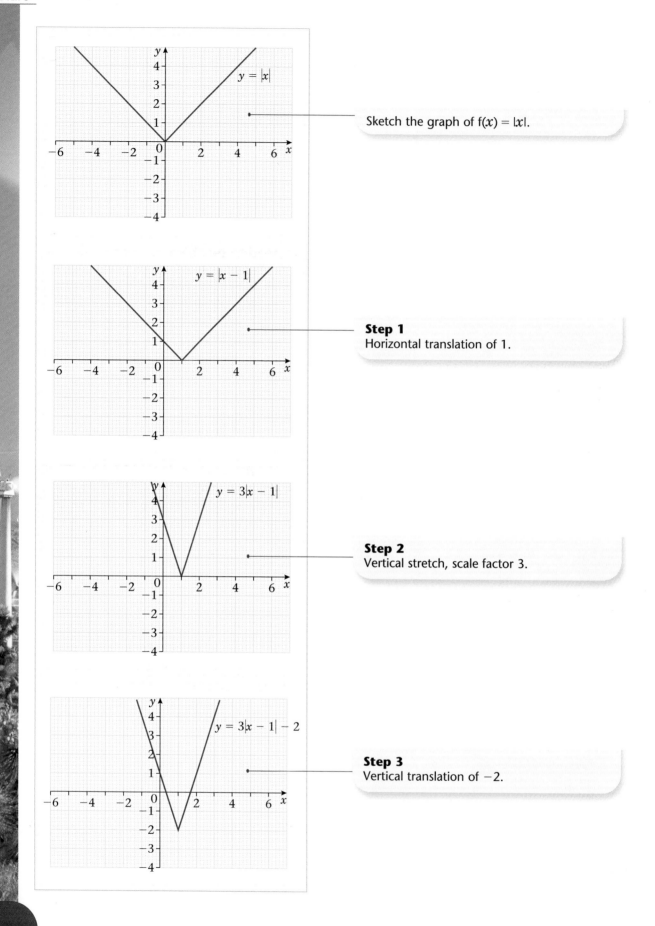

Sketch the graph of $f(x) = |x|$.

Step 1
Horizontal translation of 1.

Step 2
Vertical stretch, scale factor 3.

Step 3
Vertical translation of -2.

Exercise 5D

1 Using combinations of transformations, sketch the graph of each of the following:

a $y = 2x^2 - 4$

b $y = 3(x + 1)^2$

c $y = \dfrac{3}{x} - 2$

d $y = \dfrac{3}{x - 2}$

e $y = 5\sin(x + 30°),\ 0 \leqslant x \leqslant 360°$

f $y = \frac{1}{2}e^x + 4$

g $y = |4x| + 1$

h $y = 2x^3 - 3$

i $y = 3\ln(x - 2),\ x > 2$

j $y = |2e^x - 3|$

5.5 When you are given a sketch of $y = f(x)$, you need to be able to sketch transformations of the graph, showing coordinates of the points to which given points are mapped.

Example 15

The diagram shows a sketch of the graph of $y = f(x)$. The curve passes through the origin O, the point $A(2, -1)$ and the point $B(6, 4)$.

Sketch the graph of:

a $y = 2f(x) - 1$

b $y = f(x + 2) + 2$

c $y = \frac{1}{4}f(2x)$

d $y = -f(x - 1)$

In each case, find the coordinates of the images of the points O, A and B.

a $y = 2f(x) - 1$

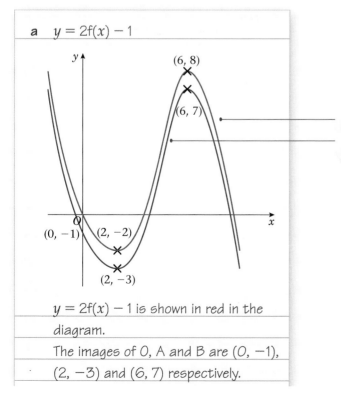

Vertical stretch, scale factor 2.

Vertical stretch, scale factor 2, then a vertical translation of -1.

$y = 2f(x) - 1$ is shown in red in the diagram.

The images of O, A and B are $(0, -1)$, $(2, -3)$ and $(6, 7)$ respectively.

b $y = f(x + 2) + 2$

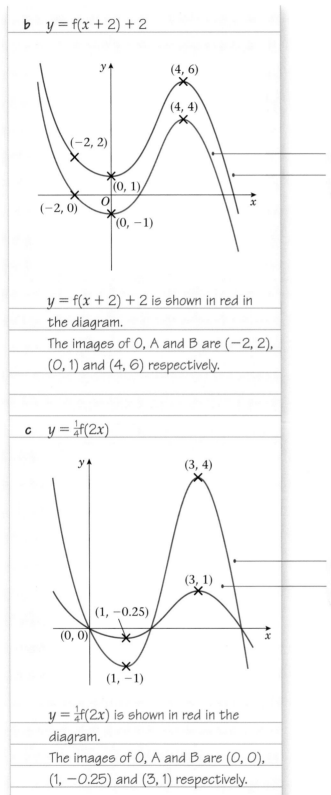

Horizontal translation of -2.

Horizontal translation of -2, then a vertical translation of 2.

$y = f(x + 2) + 2$ is shown in red in the diagram.

The images of O, A and B are $(-2, 2)$, $(0, 1)$ and $(4, 6)$ respectively.

c $y = \frac{1}{4}f(2x)$

Horizontal stretch, scale factor $\frac{1}{2}$.

Horizontal stretch, scale factor $\frac{1}{2}$, then a vertical stretch, scale factor $\frac{1}{4}$.

$y = \frac{1}{4}f(2x)$ is shown in red in the diagram.

The images of O, A and B are $(0, 0)$, $(1, -0.25)$ and $(3, 1)$ respectively.

d $y = -f(x - 1)$

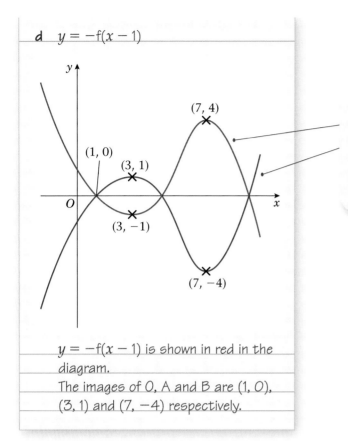

(7, 4)

(1, 0)

(3, 1)

(3, −1)

(7, −4)

Horizontal translation of 1.

Horizontal translation of 1, then a vertical stretch, scale factor −1.

A 'vertical stretch with scale factor −1' is equivalent to a reflection in the x-axis.

$y = -f(x - 1)$ is shown in red in the diagram.
The images of O, A and B are (1, 0), (3, 1) and (7, −4) respectively.

Exercise 5E

1 The diagram shows a sketch of the graph of $y = f(x)$.
The curve passes through the origin O, the point A(−2, −2) and the point B(3, 4).

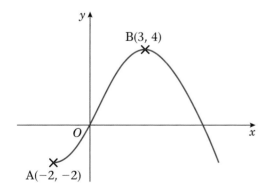

B(3, 4)

O

A(−2, −2)

Sketch the graph of:

a $y = 3f(x) + 2$

b $y = f(x - 2) - 5$

c $y = \frac{1}{2}f(x + 1)$

d $y = -f(2x)$

In each case, find the coordinates of the images of the points O, A and B.

2 The diagram shows a sketch of the graph of $y = f(x)$. The curve has a maximum at the point A(-1, 4) and crosses the axes at the points B(0, 3) and C(-2, 0).

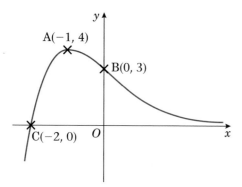

Sketch the graph of:

a $y = 3f(x - 2)$ **b** $y = \frac{1}{2}f(\frac{1}{2}x)$ **c** $y = -f(x) + 4$ **d** $y = -2f(x + 1)$

For each graph, find, where possible, the coordinates of the maximum or minimum and the coordinates of the intersection points with the axes.

3 The diagram shows a sketch of the graph of $y = f(x)$. The lines $x = 2$ and $y = 0$ (the x-axis) are asymptotes to the curve.

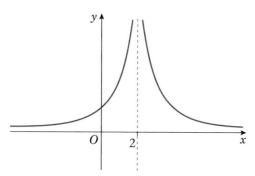

Sketch the graph of:

a $y = 3f(x) - 1$ **b** $y = f(x + 2) + 4$ **c** $y = -f(2x)$

For each part, state the equations of the asymptotes.

Mixed exercise 5F

1 a Using the same scales and the same axes, sketch the graphs of $y = |2x|$ and $y = |x - a|$, where $a > 0$.

b Write down the coordinates of the points where the graph of $y = |x - a|$ meets the axes.

c Show that the point with coordinates $(-a, 2a)$ lies on both graphs.

d Find the coordinates, in terms of a, of a second point which lies on both graphs. **E**

2 a Sketch, on a single diagram, the graphs of $y = a^2 - x^2$ and $y = |x + a|$, where a is a constant and $a > 1$.

b Write down the coordinates of the points where the graph of $y = a^2 - x^2$ cuts the coordinate axes.

c Given that the two graphs intersect at $x = 4$, calculate the value of a. **E**

3 **a** On the same axes, sketch the graphs of $y = 2 - x$ and $y = 2|x + 1|$.

   b Hence, or otherwise, find the values of x for which $2 - x = 2|x + 1|$.   **E**

4 Functions f and g are defined by

$$f : x \rightarrow 4 - x \quad \{x \in \mathbb{R}\}$$
$$g : x \rightarrow 3x^2 \quad \{x \in \mathbb{R}\}$$

   a Find the range of g.

   b Solve $gf(x) = 48$.

   c Sketch the graph of $y = |f(x)|$ and hence find the values of x for which $|f(x)| = 2$.   **E**

5 The function f is defined by $f : x \rightarrow |2x - a| \quad \{x \in \mathbb{R}\}$, where a is a positive constant.

   a Sketch the graph of $y = f(x)$, showing the coordinates of the points where the graph cuts the axes.

   b On a separate diagram, sketch the graph of $y = f(2x)$, showing the coordinates of the points where the graph cuts the axes.

   c Given that a solution of the equation $f(x) = \frac{1}{2}x$ is $x = 4$, find the two possible values of a.   **E**

6 **a** Sketch the graph of $y = |x - 2a|$, where a is a positive constant. Show the coordinates of the points where the graph meets the axes.

   b Using algebra solve, for x in terms of a, $|x - 2a| = \frac{1}{3}x$.

   c On a separate diagram, sketch the graph of $y = a - |x - 2a|$, where a is a positive constant. Show the coordinates of the points where the graph cuts the axes.   **E**

7 **a** Sketch the graph of $y = |2x + a|$, $a > 0$, showing the coordinates of the points where the graph meets the coordinate axes.

   b On the same axes, sketch the graph of $y = \dfrac{1}{x}$.

   c Explain how your graphs show that there is only one solution of the equation $x|2x + a| - 1 = 0$.

   d Find, using algebra, the value of x for which $x|2x + a| - 1 = 0$.   **E**

8 The diagram shows part of the curve with equation $y = f(x)$, where

$$f(x) = x^2 - 7x + 5 \ln x + 8 \quad x > 0$$

The points A and B are the stationary points of the curve.

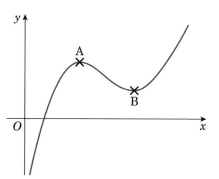

   a Using calculus and showing your working, find the coordinates of the points A and B.

   b Sketch the curve with equation $y = -3f(x - 2)$.

   c Find the coordinates of the stationary points of the curve with equation $y = -3f(x - 2)$. State, without proof, which point is a maximum and which point is a minimum.   **E**

Summary of key points

1   The modulus of a number a, written as $|a|$, is its **positive** numerical value.

- For $a \geqslant 0$, $|a| = a$.
- For $a < 0$, $|a| = -a$.

2   To sketch the graph of $y = |f(x)|$:

- Sketch the graph of $y = f(x)$.
- Reflect in the x-axis any parts where $f(x) < 0$ (parts below the x-axis).
- Delete the parts below the x-axis.

3   To sketch the graph of $y = f(|x|)$:

- Sketch the graph of $y = f(x)$ for $x \geqslant 0$.
- Reflect this in the y-axis.

4   To solve an equation of the type $|f(x)| = g(x)$ or $|f(x)| = |g(x)|$:

- Use a sketch to locate the roots.
- Solve algebraically, using $-f(x)$ for reflected parts of $y = f(x)$ and $-g(x)$ for reflected parts of $y = g(x)$.

5   Basic types of transformation are

$f(x + a)$ a horizontal translation of $-a$

$f(x) + a$ a vertical translation of $+a$

$f(ax)$ a horizontal stretch of scale factor $\dfrac{1}{a}$

$af(x)$ a vertical stretch of scale factor a

These may be combined to give, for example $bf(x + a)$, which is a horizontal translation of $-a$ followed by a vertical stretch of scale factor b.

6   For combinations of transformations, the graph can be built up 'one step at a time', starting from a basic or given curve.

After completing this chapter you should know

1 the functions secant θ, cosecant θ and cotangent θ

2 the graphs of sec θ, cosec θ and cot θ

3 how to solve equations and prove identities involving sec θ, cosec θ and cot θ

4 how to prove and use the identities
$$1 + \tan^2\theta = \sec^2\theta$$
and $1 + \cot^2\theta = \csc^2\theta$

5 how to sketch and use the inverse trigonometric functions arcsin$x$, arccos$x$ and arctan$x$.

Trigonometry

Leonhard Euler (1707–1783), the famous Swiss mathematician, is generally regarded as the man responsible for introducing the terminology and notation surrounding trigonometric functions and identities.

6.1 You need to know the functions secant θ, cosecant θ and cotangent θ.

■ The functions secant θ, cosecant θ and cotangent θ are defined as:

> These are often written and pronounced as **sec θ**, **cosec θ** and **cot θ**.

● $\sec \theta = \dfrac{1}{\cos \theta}$

{undefined for values of θ at which cos θ = 0}

> Remember that $\cos^n \theta \equiv (\cos \theta)^n$ for $n \in \mathbb{Z}^+$. The convention is not used for $n \in \mathbb{Z}^-$.
> For example, $\cos^{-1} \theta$ does not mean $\dfrac{1}{\cos \theta}$.
> Do not confuse $\cos^{-1} \theta$ with sec θ.

● $\operatorname{cosec} \theta = \dfrac{1}{\sin \theta}$

{undefined for values of θ at which sin θ = 0}

● $\cot \theta = \dfrac{1}{\tan \theta}$

{undefined for values of θ at which tan θ = 0}.

> As $\tan \theta = \dfrac{\sin \theta}{\cos \theta}$, cot θ can also be written as $\cot \theta = \dfrac{\cos \theta}{\sin \theta}$.

Example 1

Use your calculator to write down the value of:

a sec 280°

b cot 115°.

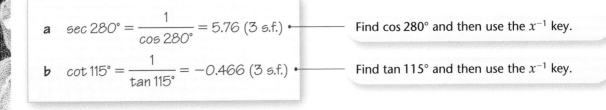

a $\sec 280° = \dfrac{1}{\cos 280°} = 5.76 \ (3 \text{ s.f.})$ ——— Find cos 280° and then use the x^{-1} key.

b $\cot 115° = \dfrac{1}{\tan 115°} = -0.466 \ (3 \text{ s.f.})$ ——— Find tan 115° and then use the x^{-1} key.

Example 2

Work out the *exact* values of:

a sec 210°

b $\operatorname{cosec} \dfrac{3\pi}{4}$.

> *Exact* here means give in surd form.

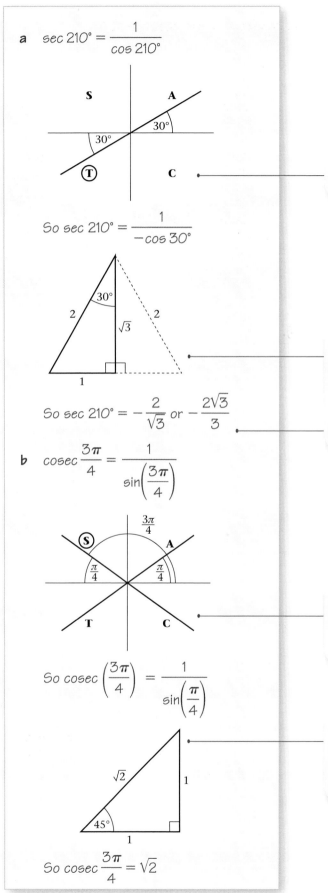

a $sec\ 210° = \dfrac{1}{cos\ 210°}$

210° is in 3rd quadrant, so $cos\ 210° = -cos\ 30°$.

So $sec\ 210° = \dfrac{1}{-cos\ 30°}$

Remember that $cos\ 30° = \dfrac{\sqrt{3}}{2}$, or draw an equilateral triangle of side 2 and use Pythagoras' theorem.

So $sec\ 210° = -\dfrac{2}{\sqrt{3}}$ or $-\dfrac{2\sqrt{3}}{3}$

Rationalise the denominator.

b $cosec\ \dfrac{3\pi}{4} = \dfrac{1}{sin\left(\dfrac{3\pi}{4}\right)}$

$\dfrac{3\pi}{4}$ (135°) is in the 2nd quadrant, so $sin\ \dfrac{3\pi}{4} = +sin\ \dfrac{\pi}{4}$.

So $cosec\left(\dfrac{3\pi}{4}\right) = \dfrac{1}{sin\left(\dfrac{\pi}{4}\right)}$

Remember that $sin\ \dfrac{\pi}{4} = \dfrac{1}{\sqrt{2}}$, or draw a right-angled isosceles triangle and use Pythagoras' theorem.

So $cosec\ \dfrac{3\pi}{4} = \sqrt{2}$

Exercise 6A

1 Without using your calculator, write down the sign of the following trigonometric ratios:

   a $\sec 300°$ +   **b** $\operatorname{cosec} 190°$ −   **c** $\cot 110°$ −

   d $\cot 200°$ −   **e** $\sec 95°$ +

2 Use your calculator to find, to 3 significant figures, the values of

   a $\sec 100°$ **b** $\operatorname{cosec} 260°$ **c** $\operatorname{cosec} 280°$

   d $\cot 550°$ **e** $\cot \dfrac{4\pi}{3}$ **f** $\sec 2.4^c$

   g $\operatorname{cosec} \dfrac{11\pi}{10}$ **h** $\sec 6^c$

3 Find the exact value (in surd form where appropriate) of the following:

   a $\operatorname{cosec} 90°$ 1   **b** $\cot 135°$ -1 **c** $\sec 180°$ -1

   d $\sec 240°$ -2 **e** $\operatorname{cosec} 300°$ $-\dfrac{2\sqrt{3}}{3}$ **f** $\cot(-45°)$ -1

   g $\sec 60°$ 2   **h** $\operatorname{cosec}(-210°)$ 2   **i** $\sec 225°$ $-\sqrt{2}$

   j $\cot \dfrac{4\pi}{3}$ $\dfrac{\sqrt{3}}{3}$ **k** $\sec \dfrac{11\pi}{6}$ $\dfrac{2\sqrt{3}}{3}$ **l** $\operatorname{cosec}\left(-\dfrac{3\pi}{4}\right)$ $-\sqrt{2}$

4 **a** Copy and complete the table, showing values (to 2 decimal places) of $\sec\theta$ for selected values of θ.

θ	0°	30°	45°	60°	70°	80°	85°	95°	100°	110°	120°	135°	150°	180°	210°
$\sec\theta$	1		1.41			5.76	11.47			−2.92		−1.41			−1.15

   b Copy and complete the table, showing values (to 2 decimal places) of $\operatorname{cosec}\theta$ for selected values of θ.

θ	10°	20°	30°	45°	60°	80°	90°	100°	120°	135°	150°	160°	170°
$\operatorname{cosec}\theta$				1.41			1		1.15	1.41			

θ	190°	200°	210°	225°	240°	270°	300°	315°	330°	340°	350°	390°
$\operatorname{cosec}\theta$				−1.15					−2			

   c Copy and complete the table, showing values (to 2 decimal places) of $\cot\theta$ for selected values of θ.

θ	−90°	−60°	−45°	−30°	−10°	10°	30°	45°	60°	90°	120°	135°	150°	170°	210°	225°	240°	270°
$\cot\theta$	0	−0.58					1.73	1	0.58			−1					0.58	

6.2 You need to know the graphs of sec θ, cosec θ and cot θ.

Example 3

Sketch, in the interval $-180° \leqslant \theta \leqslant 180°$, the graph of $y = \sec \theta$.

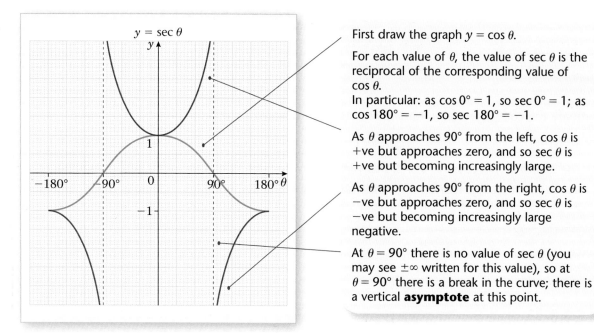

First draw the graph $y = \cos \theta$.

For each value of θ, the value of $\sec \theta$ is the reciprocal of the corresponding value of $\cos \theta$.
In particular: as $\cos 0° = 1$, so $\sec 0° = 1$; as $\cos 180° = -1$, so $\sec 180° = -1$.

As θ approaches 90° from the left, $\cos \theta$ is +ve but approaches zero, and so $\sec \theta$ is +ve but becoming increasingly large.

As θ approaches 90° from the right, $\cos \theta$ is −ve but approaches zero, and so $\sec \theta$ is −ve but becoming increasingly large negative.

At $\theta = 90°$ there is no value of $\sec \theta$ (you may see $\pm\infty$ written for this value), so at $\theta = 90°$ there is a break in the curve; there is a vertical **asymptote** at this point.

Compare the completed table for Question 4a in Exercise 6A with the related part of the graph in Example 3.

■ **The graph of $y = \sec \theta$, $\theta \in \mathbb{R}$, has symmetry in the y-axis and repeats itself every 360°. It has vertical asymptotes at all the values of θ for which $\cos \theta = 0$, i.e. at $\theta = 90° + 180n°$, $n \in \mathbb{Z}$.**

Example 4

Sketch the graph of $y = \text{cosec } \theta$.

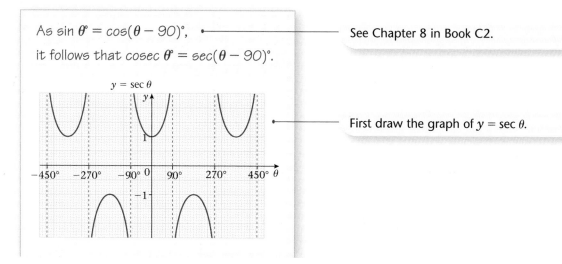

As $\sin \theta° = \cos(\theta - 90)°$,

it follows that $\text{cosec } \theta° = \sec(\theta - 90)°$.

See Chapter 8 in Book C2.

First draw the graph of $y = \sec \theta$.

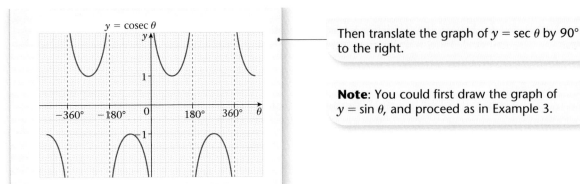

Then translate the graph of $y = \sec \theta$ by 90° to the right.

Note: You could first draw the graph of $y = \sin \theta$, and proceed as in Example 3.

Compare the completed table for Question 4b in Exercise 6A with the graph of $y = \csc \theta$ in Example 4.

■ **The graph of $y = \csc \theta$, $\theta \in \mathbb{R}$, has vertical asymptotes at all the values of θ for which $\sin \theta = 0$, i.e. at $\theta = 180n°$, $n \in \mathbb{Z}$, and the curve repeats itself every 360°.**

Example 5

Sketch the graph of $y = \cot \theta$.

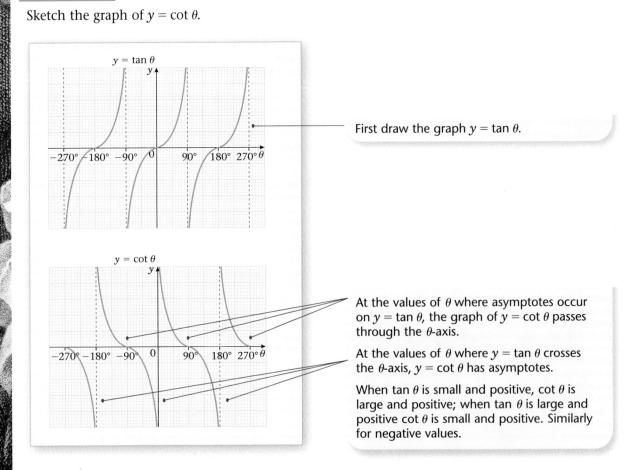

First draw the graph $y = \tan \theta$.

At the values of θ where asymptotes occur on $y = \tan \theta$, the graph of $y = \cot \theta$ passes through the θ-axis.

At the values of θ where $y = \tan \theta$ crosses the θ-axis, $y = \cot \theta$ has asymptotes.

When $\tan \theta$ is small and positive, $\cot \theta$ is large and positive; when $\tan \theta$ is large and positive $\cot \theta$ is small and positive. Similarly for negative values.

Compare the graph in Example 5 with your answers to Exercise 6A, Question 4c.

■ **The graph of $y = \cot \theta$, $\theta \in \mathbb{R}$, has vertical asymptotes at all the values of θ for which $\sin \theta = 0$, i.e. at $\theta = 180n°$, $n \in \mathbb{Z}$, and the curve repeats itself every 180°.**

Example 6

Sketch, in the interval $0 \le \theta \le 360°$, the graph of $y = 1 + \sec 2\theta$.

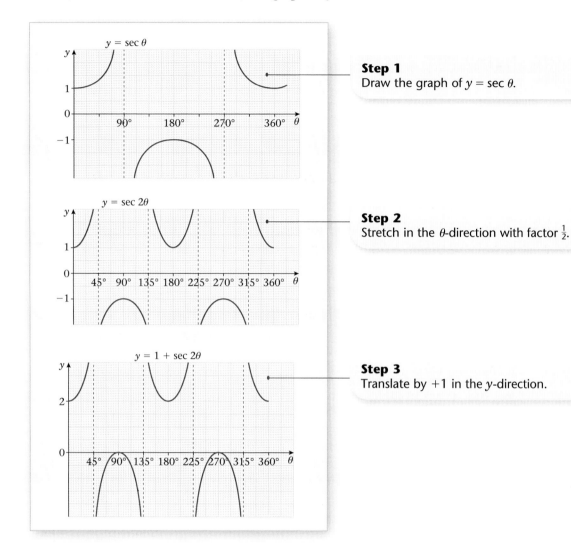

Step 1
Draw the graph of $y = \sec \theta$.

Step 2
Stretch in the θ-direction with factor $\frac{1}{2}$.

Step 3
Translate by $+1$ in the y-direction.

Exercise 6B

1 a Sketch, in the interval $-540° \le \theta \le 540°$, the graphs of:
   i $\sec \theta$ **ii** $\operatorname{cosec} \theta$ **iii** $\cot \theta$

b Write down the range of
   i $\sec \theta$ **ii** $\operatorname{cosec} \theta$ **iii** $\cot \theta$

2 a Sketch, on the same set of axes, in the interval $0 \le \theta \le 360°$, the graphs of $y = \sec \theta$ and $y = -\cos \theta$.

b Explain how your graphs show that $\sec \theta = -\cos \theta$ has no solutions.

3 a Sketch, on the same set of axes, in the interval $0 \le \theta \le 360°$, the graphs of $y = \cot \theta$ and $y = \sin 2\theta$.

b Deduce the number of solutions of the equation $\cot \theta = \sin 2\theta$ in the interval $0 \le \theta \le 360°$.

4 a Sketch on separate axes, in the interval $0 \leqslant \theta \leqslant 360°$, the graphs of $y = \tan \theta$ and $y = \cot(\theta + 90°)$.

b Hence, state a relationship between $\tan \theta$ and $\cot(\theta + 90°)$.

5 a Describe the relationships between the graphs of

i $\tan\left(\theta + \dfrac{\pi}{2}\right)$ and $\tan \theta$ **ii** $\cot(-\theta)$ and $\cot \theta$

iii $\operatorname{cosec}\left(\theta + \dfrac{\pi}{4}\right)$ and $\operatorname{cosec} \theta$ **iv** $\sec\left(\theta - \dfrac{\pi}{4}\right)$ and $\sec \theta$

b By considering the graphs of $\tan\left(\theta + \dfrac{\pi}{2}\right)$, $\cot(-\theta)$, $\operatorname{cosec}\left(\theta + \dfrac{\pi}{4}\right)$ and $\sec\left(\theta - \dfrac{\pi}{4}\right)$, state which pairs of functions are equal.

6 Sketch on separate axes, in the interval $0 \leqslant \theta \leqslant 360°$, the graphs of:

a $y = \sec 2\theta$ **b** $y = -\operatorname{cosec} \theta$ **c** $y = 1 + \sec \theta$ **d** $y = \operatorname{cosec}(\theta - 30°)$

In each case show the coordinates of any maximum and minimum points, and of any points at which the curve meets the axes.

7 Write down the periods of the following functions. Give your answer in terms of π.

a $\sec 3\theta$ **b** $\operatorname{cosec} \tfrac{1}{2}\theta$ **c** $2 \cot \theta$ **d** $\sec(-\theta)$

8 a Sketch the graph of $y = 1 + 2 \sec \theta$ in the interval $-\pi \leqslant \theta \leqslant 2\pi$.

b Write down the y-coordinate of points at which the gradient is zero.

c Deduce the maximum and minimum values of $\dfrac{1}{1 + 2 \sec \theta}$, and give the smallest positive values of θ at which they occur.

6.3 **You need to be able to simplify expressions, prove identities and solve equations involving secant θ, cosecant θ and cotangent θ.**

Example 7

Simplify

a $\sin \theta \cot \theta \sec \theta$ **b** $\sin \theta \cos \theta(\sec \theta + \operatorname{cosec} \theta)$

a $\sin \theta \cot \theta \sec \theta$

$= \overset{1}{\cancel{\sin \theta}} \times \dfrac{\cos \theta}{\overset{}{\cancel{\sin \theta}}_{1}} \times \dfrac{1}{\overset{}{\cancel{\cos \theta}}^{1}}$

$= 1$

Write the expression in terms of sin and cos, using $\cot \theta = \dfrac{\cos \theta}{\sin \theta}$ and $\sec \theta = \dfrac{1}{\cos \theta}$.

b $\sec \theta + \operatorname{cosec} \theta = \dfrac{1}{\cos \theta} + \dfrac{1}{\sin \theta}$

$\quad\quad\quad = \dfrac{\sin \theta + \cos \theta}{\sin \theta \cos \theta}$

So $\sin \theta \cos \theta(\sec \theta + \operatorname{cosec} \theta)$
$= \sin \theta + \cos \theta$

The given expression reduces to
$\sin \theta + \cos \theta$.

Write the expression in terms of sin and cos, using $\sec \theta = \dfrac{1}{\cos \theta}$ and $\operatorname{cosec} \theta = \dfrac{1}{\sin \theta}$.

Put over common denominator.

Multiply both sides by $\sin \theta \cos \theta$.

Example 8

Show that $\dfrac{\cot\theta\cosec\theta}{\sec^2\theta + \cosec^2\theta} \equiv \cos^3\theta$

Consider LHS:

The numerator $\cot\theta\cosec\theta$

$$\equiv \frac{\cos\theta}{\sin\theta} \times \frac{1}{\sin\theta} \equiv \frac{\cos\theta}{\sin^2\theta}$$

Write the expression in terms of sin and cos, using $\cot\theta = \dfrac{\cos\theta}{\sin\theta}$ and $\cosec\theta = \dfrac{1}{\sin\theta}$.

The denominator $\sec^2\theta + \cosec^2\theta$

$$\equiv \frac{1}{\cos^2\theta} + \frac{1}{\sin^2\theta}$$

$$\equiv \frac{\sin^2\theta + \cos^2\theta}{\cos^2\theta\sin^2\theta}$$

$$\equiv \frac{1}{\cos^2\theta\sin^2\theta}$$

Write the expression in terms of sin and cos, using $\sec^2\theta = \left(\dfrac{1}{\cos\theta}\right)^2 \equiv \dfrac{1}{\cos^2\theta}$ and $\cosec^2\theta \equiv \dfrac{1}{\sin^2\theta}$.

Remember that $\sin^2\theta + \cos^2\theta \equiv 1$.

So $\dfrac{\cot\theta\cosec\theta}{\sec^2\theta + \cosec^2\theta}$

$$\equiv \left(\frac{\cos\theta}{\sin^2\theta}\right) \div \left(\frac{1}{\cos^2\theta\sin^2\theta}\right)$$

$$\equiv \frac{\cos\theta}{\sin^2\theta} \times \frac{\cos^2\theta\sin^2\theta}{1}$$

Remember to invert the fraction when changing from ÷ sign to ×.

$$\equiv \cos^3\theta$$

Example 9

Solve the equations:

a $\sec\theta = -2.5$ **b** $\cot 2\theta = 0.6$

in the interval $0 \leqslant \theta \leqslant 360°$.

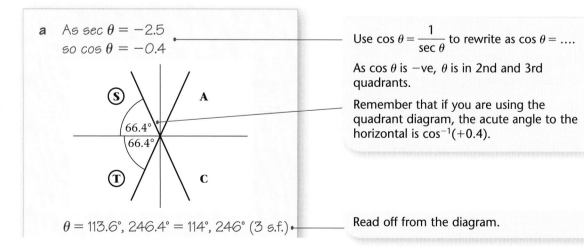

a As $\sec\theta = -2.5$

so $\cos\theta = -0.4$

Use $\cos\theta = \dfrac{1}{\sec\theta}$ to rewrite as $\cos\theta = ...$

As $\cos\theta$ is −ve, θ is in 2nd and 3rd quadrants.

Remember that if you are using the quadrant diagram, the acute angle to the horizontal is $\cos^{-1}(+0.4)$.

$\theta = 113.6°, 246.4° = 114°, 246°$ (3 s.f.)

Read off from the diagram.

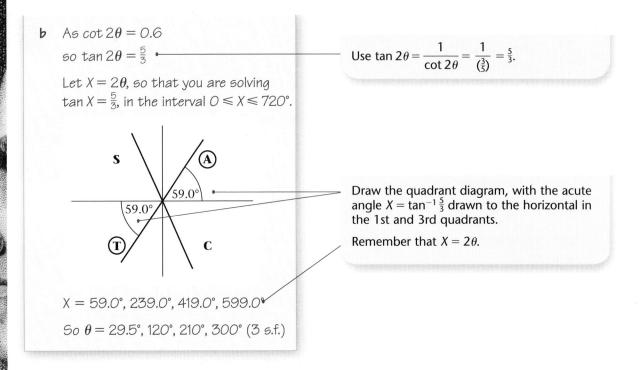

b As $\cot 2\theta = 0.6$

so $\tan 2\theta = \frac{5}{3}$

Use $\tan 2\theta = \dfrac{1}{\cot 2\theta} = \dfrac{1}{\left(\frac{3}{5}\right)} = \frac{5}{3}$.

Let $X = 2\theta$, so that you are solving $\tan X = \frac{5}{3}$, in the interval $0 \leqslant X \leqslant 720°$.

Draw the quadrant diagram, with the acute angle $X = \tan^{-1}\frac{5}{3}$ drawn to the horizontal in the 1st and 3rd quadrants.

Remember that $X = 2\theta$.

$X = 59.0°, 239.0°, 419.0°, 599.0°$

So $\theta = 29.5°, 120°, 210°, 300°$ (3 s.f.)

Exercise 6C

Give solutions to these equations, correct to 1 decimal place.

1 Rewrite the following as powers of sec θ, cosec θ or cot θ:

a $\dfrac{1}{\sin^3 \theta}$ 　　　 **b** $\sqrt{\dfrac{4}{\tan^6 \theta}}$ 　　　 **c** $\dfrac{1}{2\cos^2 \theta}$ 　　　 **d** $\dfrac{1 - \sin^2 \theta}{\sin^2 \theta}$

e $\dfrac{\sec \theta}{\cos^4 \theta}$ 　　 **f** $\sqrt{\operatorname{cosec}^3 \theta \cot \theta \sec \theta}$ 　**g** $\dfrac{2}{\sqrt{\tan \theta}}$ 　　**h** $\dfrac{\operatorname{cosec}^2 \theta \tan^2 \theta}{\cos \theta}$

2 Write down the value(s) of $\cot x$ in each of the following equations:

a $5\sin x = 4\cos x$ 　　　 **b** $\tan x = -2$ 　　　 **c** $3\dfrac{\sin x}{\cos x} = \dfrac{\cos x}{\sin x}$

3 Using the definitions of **sec**, **cosec**, **cot** and **tan** simplify the following expressions:

a $\sin \theta \cot \theta$ 　　　　　　　　　　 **b** $\tan \theta \cot \theta$

c $\tan 2\theta \operatorname{cosec} 2\theta$ 　　　　　　　 **d** $\cos \theta \sin \theta (\cot \theta + \tan \theta)$

e $\sin^3 x \operatorname{cosec} x + \cos^3 x \sec x$ 　 **f** $\sec A - \sec A \sin^2 A$ $\dfrac{1}{\cos} = \dfrac{1}{\cos} \times \sin^2$

$\dfrac{1-\sin^2}{\cos} = \dfrac{\sin}{\sin} \times \tan$ 　 $\cos$

g $\sec^2 x \cos^5 x + \cot x \operatorname{cosec} x \sin^4 x$

$\dfrac{1}{\cos^2} \times \cos^5 + \dfrac{\cos}{\sin} \times \dfrac{1}{\sin} \times \sin^4$

4 Show that 　$\cos^3 +$ 　　　 $\cos\sin^2$ 　$\cos$

a $\cos \theta + \sin \theta \tan \theta \equiv \sec \theta$ 　　 **b** $\cot \theta + \tan \theta \equiv \operatorname{cosec} \theta \sec \theta$

c $\operatorname{cosec} \theta - \sin \theta \equiv \cos \theta \cot \theta$ 　 **d** $(1 - \cos x)(1 + \sec x) \equiv \sin x \tan x$

e $\dfrac{\cos x}{1 - \sin x} + \dfrac{1 - \sin x}{\cos x} \equiv 2\sec x$ 　 **f** $\dfrac{\cos \theta}{1 + \cot \theta} \equiv \dfrac{\sin \theta}{1 + \tan \theta}$

5 Solve, for values of θ in the interval $0 \leqslant \theta \leqslant 360°$, the following equations. Give your answers to 3 significant figures where necessary.

    a $\sec\theta = \sqrt{2}$ **b** $\operatorname{cosec}\theta = -3$ **c** $5\cot\theta = -2$ **d** $\operatorname{cosec}\theta = 2$

    e $3\sec^2\theta - 4 = 0$ **f** $5\cos\theta = 3\cot\theta$ **g** $\cot^2\theta - 8\tan\theta = 0$ **h** $2\sin\theta = \operatorname{cosec}\theta$

6 Solve, for values of θ in the interval $-180° \leqslant \theta \leqslant 180°$, the following equations:

    a $\operatorname{cosec}\theta = 1$ **b** $\sec\theta = -3$

    c $\cot\theta = 3.45$ **d** $2\operatorname{cosec}^2\theta - 3\operatorname{cosec}\theta = 0$

    e $\sec\theta = 2\cos\theta$ **f** $3\cot\theta = 2\sin\theta$

    g $\operatorname{cosec} 2\theta = 4$ **h** $2\cot^2\theta - \cot\theta - 5 = 0$

7 Solve the following equations for values of θ in the interval $0 \leqslant \theta \leqslant 2\pi$. Give your answers in terms of π.

    a $\sec\theta = -1$ **b** $\cot\theta = -\sqrt{3}$

    c $\operatorname{cosec}\frac{1}{2}\theta = \dfrac{2\sqrt{3}}{3}$ **d** $\sec\theta = \sqrt{2}\tan\theta \left(\theta \neq \dfrac{\pi}{2},\ \theta \neq \dfrac{3\pi}{2}\right)$

8 In the diagram $AB = 6\,\text{cm}$ is the diameter of the circle and BT is the tangent to the circle at B. The chord AC is extended to meet this tangent at D and $\angle DAB = \theta$.

    a Show that $CD = 6(\sec\theta - \cos\theta)$.

    b Given that $CD = 16\,\text{cm}$, calculate the length of the chord AC.

6.4 **You need to know and be able to use the identities**
    ● $1 + \tan^2\theta \equiv \sec^2\theta$
    ● $1 + \cot^2\theta \equiv \operatorname{cosec}^2\theta$

Example **10**

Show that $1 + \tan^2\theta \equiv \sec^2\theta$

As $\sin^2\theta + \cos^2\theta \equiv 1$

so $\dfrac{\sin^2\theta}{\cos^2\theta} + \dfrac{\cos^2\theta}{\cos^2\theta} \equiv \dfrac{1}{\cos^2\theta}$ Divide both sides of the identity by $\cos^2\theta$.

so $\left(\dfrac{\sin\theta}{\cos\theta}\right)^2 + 1 \equiv \left(\dfrac{1}{\cos\theta}\right)^2$ Use $\tan\theta = \dfrac{\sin\theta}{\cos\theta}$ and $\sec\theta = \dfrac{1}{\cos\theta}$

∴ $1 + \tan^2\theta \equiv \sec^2\theta$

Example 11

Show that $1 + \cot^2 \theta \equiv \csc^2 \theta$

As $\quad \sin^2 \theta + \cos^2 \theta \equiv 1$

so $\quad \dfrac{\sin^2 \theta}{\sin^2 \theta} + \dfrac{\cos^2 \theta}{\sin^2 \theta} \equiv \dfrac{1}{\sin^2 \theta}$ •————— Divide both sides of the identity by $\sin^2 \theta$.

so $\quad 1 + \left(\dfrac{\cos \theta}{\sin \theta}\right)^2 \equiv \left(\dfrac{1}{\sin \theta}\right)^2$ •————— Use $\cot \theta = \dfrac{\cos \theta}{\sin \theta}$ and $\csc \theta = \dfrac{1}{\sin \theta}$

$\therefore \quad 1 + \cot^2 \theta \equiv \csc^2 \theta$

Example 12

Given that $\tan A = -\frac{5}{12}$, and that angle A is obtuse, find the exact value of

a $\sec A$ **b** $\sin A$

a **Method 1**
Using $1 + \tan^2 A \equiv \sec^2 A$
$\sec^2 A = 1 + \frac{25}{144} = \frac{169}{144}$ $\qquad$ $\tan^2 A = \frac{25}{144}$

$\sec A = \pm\frac{13}{12}$ •————— This does not take account of the fact that angle A is obtuse.

As angle A is obtuse, i.e. in the 2nd quadrant, $\sec A$ is −ve.

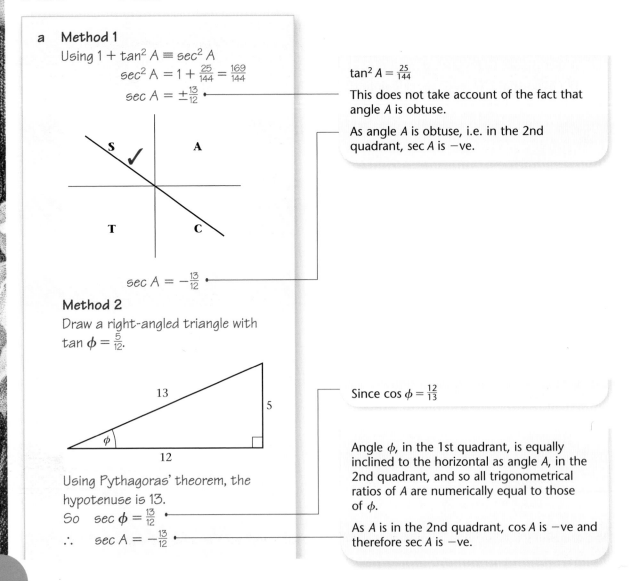

$\sec A = -\frac{13}{12}$

Method 2
Draw a right-angled triangle with $\tan \phi = \frac{5}{12}$.

Since $\cos \phi = \frac{12}{13}$

Angle ϕ, in the 1st quadrant, is equally inclined to the horizontal as angle A, in the 2nd quadrant, and so all trigonometrical ratios of A are numerically equal to those of ϕ.

Using Pythagoras' theorem, the hypotenuse is 13.
So $\quad \sec \phi = \frac{13}{12}$ •

$\therefore \quad \sec A = -\frac{13}{12}$ •————— As A is in the 2nd quadrant, $\cos A$ is −ve and therefore $\sec A$ is −ve.

b Using $\tan A \equiv \dfrac{\sin A}{\cos A}$

$\qquad \sin A \equiv \tan A \cos A$

So $\quad \sin A \equiv \left(-\frac{5}{12}\right) \times \left(-\frac{12}{13}\right)$

$\qquad\qquad \equiv \frac{5}{13}$

$\cos A = -\frac{12}{13}$, since $\cos A \equiv \dfrac{1}{\sec A}$

Example 13

Prove the identities

a $\operatorname{cosec}^4 \theta - \cot^4 \theta \equiv \dfrac{1 + \cos^2 \theta}{1 - \cos^2 \theta}$

b $\sec^2 \theta - \cos^2 \theta \equiv \sin^2 \theta (1 + \sec^2 \theta)$

a $\text{LHS} = \operatorname{cosec}^4 \theta - \cot^4 \theta$

This is the difference of two squares, so factorise.

$\equiv (\operatorname{cosec}^2 \theta + \cot^2 \theta)(\operatorname{cosec}^2 \theta - \cot^2 \theta)$

$\equiv \operatorname{cosec}^2 \theta + \cot^2 \theta$

As $1 + \cot^2 \theta \equiv \operatorname{cosec}^2 \theta$, so $\operatorname{cosec}^2 \theta - \cot^2 \theta \equiv 1$.

$\equiv \dfrac{1}{\sin^2 \theta} + \dfrac{\cos^2 \theta}{\sin^2 \theta}$

$\equiv \dfrac{1 + \cos^2 \theta}{\sin^2 \theta}$

Using $\operatorname{cosec} \theta \equiv \dfrac{1}{\sin \theta}$, $\cot \theta \equiv \dfrac{\cos \theta}{\sin \theta}$.

$\equiv \dfrac{1 + \cos^2 \theta}{1 - \cos^2 \theta} = \text{RHS}$

Using $\sin^2 \theta + \cos^2 \theta \equiv 1$.

b $\text{RHS} = \sin^2 \theta + \sin^2 \theta \sec^2 \theta$

Write in terms of $\sin \theta$ and $\cos \theta$.

$\equiv \sin^2 \theta + \dfrac{\sin^2 \theta}{\cos^2 \theta}$

Use $\sec \theta = \dfrac{1}{\cos \theta}$

$\equiv \sin^2 \theta + \tan^2 \theta$

$\dfrac{\sin^2 \theta}{\cos^2 \theta} = \left(\dfrac{\sin \theta}{\cos \theta}\right)^2 = \tan^2 \theta.$

$\equiv (1 - \cos^2 \theta) + (\sec^2 \theta - 1)$

Look at LHS. It is in terms of $\cos^2 \theta$ and $\sec^2 \theta$, so use $\sin^2 \theta + \cos^2 \theta \equiv 1$ and $1 + \tan^2 \theta \equiv \sec^2 \theta$.

$\equiv \sec^2 \theta - \cos^2 \theta$

$= \text{LHS}$

Note: Try starting with the LHS, using $\cos^2 \theta \equiv 1 - \sin^2 \theta$ and $\sec^2 \theta \equiv 1 + \tan^2 \theta$.

The identities $1 + \tan^2 \theta \equiv \sec^2 \theta$ and $1 + \cot^2 \theta \equiv \operatorname{cosec}^2 \theta$ extend the range of equations that can be solved.

Example 14

Solve the equation $4 \operatorname{cosec}^2 \theta - 9 = \cot \theta$, in the interval $0 \leqslant \theta \leqslant 360°$.

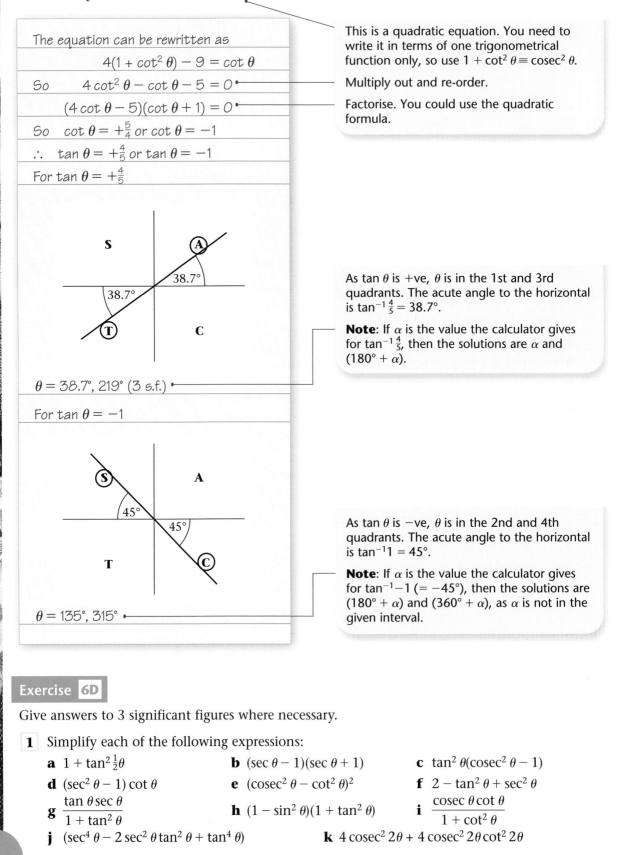

The equation can be rewritten as

$$4(1 + \cot^2 \theta) - 9 = \cot \theta$$

So $4 \cot^2 \theta - \cot \theta - 5 = 0$

$(4 \cot \theta - 5)(\cot \theta + 1) = 0$

So $\cot \theta = +\frac{5}{4}$ or $\cot \theta = -1$

∴ $\tan \theta = +\frac{4}{5}$ or $\tan \theta = -1$

For $\tan \theta = +\frac{4}{5}$

$\theta = 38.7°, 219°$ (3 s.f.)

For $\tan \theta = -1$

$\theta = 135°, 315°$

This is a quadratic equation. You need to write it in terms of one trigonometrical function only, so use $1 + \cot^2 \theta \equiv \operatorname{cosec}^2 \theta$.

Multiply out and re-order.

Factorise. You could use the quadratic formula.

As $\tan \theta$ is +ve, θ is in the 1st and 3rd quadrants. The acute angle to the horizontal is $\tan^{-1} \frac{4}{5} = 38.7°$.

Note: If α is the value the calculator gives for $\tan^{-1} \frac{4}{5}$, then the solutions are α and $(180° + \alpha)$.

As $\tan \theta$ is −ve, θ is in the 2nd and 4th quadrants. The acute angle to the horizontal is $\tan^{-1} 1 = 45°$.

Note: If α is the value the calculator gives for $\tan^{-1} -1$ ($= -45°$), then the solutions are $(180° + \alpha)$ and $(360° + \alpha)$, as α is not in the given interval.

Exercise 6D

Give answers to 3 significant figures where necessary.

1 Simplify each of the following expressions:

a $1 + \tan^2 \frac{1}{2}\theta$

b $(\sec \theta - 1)(\sec \theta + 1)$

c $\tan^2 \theta(\operatorname{cosec}^2 \theta - 1)$

d $(\sec^2 \theta - 1) \cot \theta$

e $(\operatorname{cosec}^2 \theta - \cot^2 \theta)^2$

f $2 - \tan^2 \theta + \sec^2 \theta$

g $\dfrac{\tan \theta \sec \theta}{1 + \tan^2 \theta}$

h $(1 - \sin^2 \theta)(1 + \tan^2 \theta)$

i $\dfrac{\operatorname{cosec} \theta \cot \theta}{1 + \cot^2 \theta}$

j $(\sec^4 \theta - 2 \sec^2 \theta \tan^2 \theta + \tan^4 \theta)$

k $4 \operatorname{cosec}^2 2\theta + 4 \operatorname{cosec}^2 2\theta \cot^2 2\theta$

2 Given that $\operatorname{cosec} x = \dfrac{k}{\operatorname{cosec} x}$, where $k > 1$, find, in terms of k, possible values of $\cot x$.

3 Given that $\cot \theta = -\sqrt{3}$, and that $90° < \theta < 180°$, find the exact value of

  a $\sin \theta$ $\sin = \frac{1}{2}$ **b** $\cos \theta$ $1 + \frac{1}{3} = \frac{1}{\cos^2}$ $\cos =$

4 Given that $\tan \theta = \frac{3}{4}$, and that $180° < \theta < 270°$, find the exact value of

  a $\sec \theta$ **b** $\cos \theta$ **c** $\sin \theta$

5 Given that $\cos \theta = \frac{24}{25}$, and that θ is a reflex angle, find the exact value of

  a $\tan \theta$ **b** $\operatorname{cosec} \theta$

6 Prove the following identities:

  a $\sec^4 \theta - \tan^4 \theta \equiv \sec^2 \theta + \tan^2 \theta$ **b** $\operatorname{cosec}^2 x - \sin^2 x \equiv \cot^2 x + \cos^2 x$

  c $\sec^2 A(\cot^2 A - \cos^2 A) \equiv \cot^2 A$ **d** $1 - \cos^2 \theta \equiv (\sec^2 \theta - 1)(1 - \sin^2 \theta)$

  e $\dfrac{1 - \tan^2 A}{1 + \tan^2 A} \equiv 1 - 2\sin^2 A$ **f** $\sec^2 \theta + \operatorname{cosec}^2 \theta \equiv \sec^2 \theta \operatorname{cosec}^2 \theta$

  g $\operatorname{cosec} A \sec^2 A \equiv \operatorname{cosec} A + \tan A \sec A$ **h** $(\sec \theta - \sin \theta)(\sec \theta + \sin \theta) \equiv \tan^2 \theta + \cos^2 \theta$

7 Given that $3\tan^2 \theta + 4\sec^2 \theta = 5$, and that θ is obtuse, find the exact value of $\sin \theta$.

8 Solve the following equations in the given intervals:

  a $\sec^2 \theta = 3\tan \theta$, $0 \leqslant \theta \leqslant 360°$

  b $\tan^2 \theta - 2\sec \theta + 1 = 0$, $-\pi \leqslant \theta \leqslant \pi$

  c $\operatorname{cosec}^2 \theta + 1 = 3\cot \theta$, $-180° \leqslant \theta \leqslant 180°$

  d $\cot \theta = 1 - \operatorname{cosec}^2 \theta$, $0 \leqslant \theta \leqslant 2\pi$

  e $3\sec \frac{1}{2}\theta = 2\tan^2 \frac{1}{2}\theta$, $0 \leqslant \theta \leqslant 360°$

  f $(\sec \theta - \cos \theta)^2 = \tan \theta - \sin^2 \theta$, $0 \leqslant \theta \leqslant \pi$

  g $\tan^2 2\theta = \sec 2\theta - 1$, $0 \leqslant \theta \leqslant 180°$

  h $\sec^2 \theta - (1 + \sqrt{3})\tan \theta + \sqrt{3} = 1$, $0 \leqslant \theta \leqslant 2\pi$

9 Given that $\tan^2 k = 2\sec k$,

  a find the value of $\sec k$

  b deduce that $\cos k = \sqrt{2} - 1$

  c hence solve, in the interval $0 \leqslant k \leqslant 360°$, $\tan^2 k = 2\sec k$, giving your answers to 1 decimal place.

10 Given that $a = 4\sec x$, $b = \cos x$ and $c = \cot x$,

  a express b in terms of a

  b show that $c^2 = \dfrac{16}{a^2 - 16}$

11 Given that $x = \sec \theta + \tan \theta$,

  a show that $\dfrac{1}{x} = \sec \theta - \tan \theta$.

  b Hence express $x^2 + \dfrac{1}{x^2} + 2$ in terms of θ, in its simplest form.

12 Given that $2\sec^2 \theta - \tan^2 \theta = p$ show that $\operatorname{cosec}^2 \theta = \dfrac{p-1}{p-2}$, $p \neq 2$.

6.5 You need to be able to use the inverse trigonometric functions, arcsin x, arccos x and arctan x and their graphs.

For a one-to-one function you can draw the graph of its inverse by reflecting the graph of the function in the line $y = x$. The three trigonometric functions $\sin x$, $\cos x$ and $\tan x$ only have inverse functions if their domains are restricted so that they are one-to-one functions. The notations used for these inverse functions are $\arcsin x$, $\arccos x$ and $\arctan x$ respectively ($\sin^{-1} x$, $\cos^{-1} x$ and $\tan^{-1} x$ are also used).

See Chapter 2.

Example 15

Sketch the graph of $y = \arcsin x$.

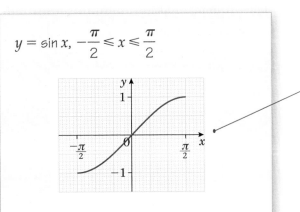

$y = \sin x, \ -\dfrac{\pi}{2} \leqslant x \leqslant \dfrac{\pi}{2}$

Step 1
Draw the graph of $y = \sin x$, with the restricted domain of $-\dfrac{\pi}{2} \leqslant x \leqslant \dfrac{\pi}{2}$

This is a **one-to-one** function, taking all

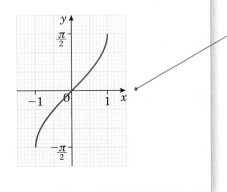

$y = \arcsin x$

Step 2
Reflect in the line $y = x$.

The domain of $\arcsin x$ is $-1 \leqslant x \leqslant 1$;
the range is $-\dfrac{\pi}{2} \leqslant \arcsin x \leqslant \dfrac{\pi}{2}$

Remember that the x and y coordinates of points interchange when reflecting in $y = x$. For example:

$$\left(\dfrac{\pi}{2}, 0\right) \rightarrow \left(0, \dfrac{\pi}{2}\right), \ (0, 1) \rightarrow (1, 0)$$

■ $\arcsin x$ is the angle α, in the interval $-\dfrac{\pi}{2} \leqslant \alpha \leqslant \dfrac{\pi}{2}$, for which $\sin \alpha = x$.

Example 16

Sketch the graph of $y = \arccos x$.

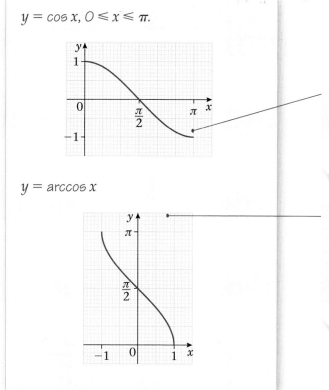

$y = \cos x, \ 0 \le x \le \pi$.

$y = \arccos x$

Step 1

Draw the graph of $y = \cos x$, with the restricted domain of $0 \le x \le \pi$.

This is a **one-to-one** function, taking all values in the range $-1 \le \cos x \le 1$.

Step 2

Reflect in the line $y = x$.

The domain of $\arccos x$ is $-1 \le x \le 1$; the range is $0 \le \arccos x \le \pi$.

Note: $(0, 1) \to (1, 0)$, $\left(\dfrac{\pi}{2}, \ 0\right) \to \left(0, \ \dfrac{\pi}{2}\right)$, $(\pi, -1) \to (-1, \pi)$.

■ $\arccos x$ is the angle α, in the interval $0 \le \alpha \le \pi$, for which $\cos \alpha = x$.

Example 17

Sketch the graph of $y = \arctan x$.

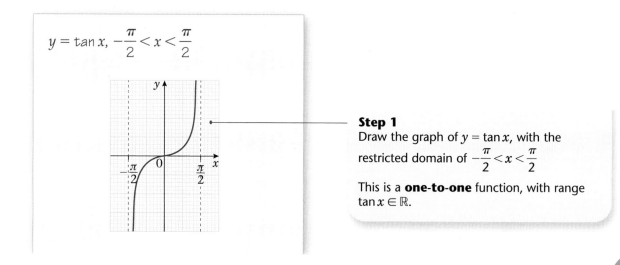

$y = \tan x, \ -\dfrac{\pi}{2} < x < \dfrac{\pi}{2}$

Step 1

Draw the graph of $y = \tan x$, with the restricted domain of $-\dfrac{\pi}{2} < x < \dfrac{\pi}{2}$

This is a **one-to-one** function, with range $\tan x \in \mathbb{R}$.

$y = \arctan x$

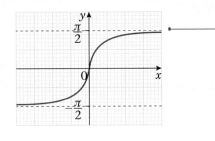

Step 2
Reflect in the line $y = x$.

The domain of $\arctan x$ is $x \in \mathbb{R}$; the range is
$-\dfrac{\pi}{2} < \arctan x < \dfrac{\pi}{2}$

■ $\arctan x$ is the angle α, in the interval $-\dfrac{\pi}{2} < \alpha < \dfrac{\pi}{2}$, for which $\tan \alpha = x$.

Example 18

Work out, in radians, the values of

a $\arcsin\left(-\dfrac{\sqrt{2}}{2}\right)$

b $\arccos(-1)$

c $\arctan(\sqrt{3})$

a

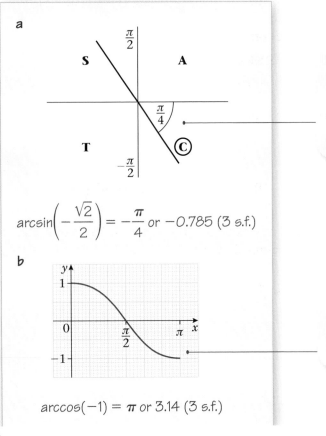

You need to solve, in the interval
$-\dfrac{\pi}{2} \leqslant x \leqslant \dfrac{\pi}{2}$, the equation $\sin x = -\dfrac{\sqrt{2}}{2}$.

The angle to the horizontal is $\dfrac{\pi}{4}$ and, as $\sin$ is $-$ve, it is in the 4th quadrant.

$\arcsin\left(-\dfrac{\sqrt{2}}{2}\right) = -\dfrac{\pi}{4}$ or -0.785 (3 s.f.)

b

You need to solve, in the interval $0 \leqslant x \leqslant \pi$, the equation $\cos x = -1$.

Draw the graph of $y = \cos x$.

$\arccos(-1) = \pi$ or 3.14 (3 s.f.)

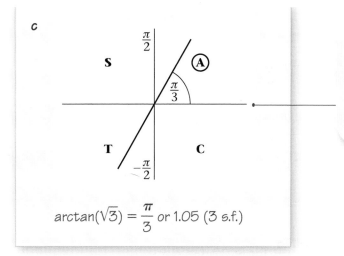

c

$$\arctan(\sqrt{3}) = \frac{\pi}{3} \text{ or } 1.05 \text{ (3 s.f.)}$$

You need to solve, in the interval $-\frac{\pi}{2} < x < \frac{\pi}{2}$, the equation $\tan x = \sqrt{3}$.

The angle to the horizontal is $\frac{\pi}{3}$ and, as tan is +ve, it is in the 1st quadrant.

Exercise 6E

1 Without using a calculator, work out, giving your answer in terms of π, the value of:

a $\arccos 0$ **b** $\arcsin(1)$ **c** $\arctan(-1)$ **d** $\arcsin(-\frac{1}{2})$

e $\arccos\left(-\frac{1}{\sqrt{2}}\right)$ **f** $\arctan\left(-\frac{1}{\sqrt{3}}\right)$ **g** $\arcsin\left(\sin\frac{\pi}{3}\right)$ **h** $\arcsin\left(\sin\frac{2\pi}{3}\right)$

2 Find the value of:

a $\arcsin(\frac{1}{2}) + \arcsin(-\frac{1}{2})$ **b** $\arccos(\frac{1}{2}) - \arccos(-\frac{1}{2})$ **c** $\arctan(1) - \arctan(-1)$

3 Without using a calculator, work out the values of:

a $\sin(\arcsin \frac{1}{2})$ **b** $\sin[\arcsin(-\frac{1}{2})]$

c $\tan[\arctan(-1)]$ **d** $\cos(\arccos 0)$

4 Without using a calculator, work out the exact values of:

a $\sin[\arccos(\frac{1}{2})]$ **b** $\cos[\arcsin(-\frac{1}{2})]$ **c** $\tan\left[\arccos\left(-\frac{\sqrt{2}}{2}\right)\right]$

d $\sec[\arctan(\sqrt{3})]$ **e** $\operatorname{cosec}[\arcsin(-1)]$ **f** $\sin\left[2\arcsin\left(\frac{\sqrt{2}}{2}\right)\right]$

5 Given that $\arcsin k = \alpha$, where $0 < k < 1$ and α is in radians, write down, in terms of α, the first two positive values of x satisfying the equation $\sin x = k$.

6 Given that x satisfies $\arcsin x = k$, where $0 < k < \frac{\pi}{2}$,

a state the range of possible values of x

b express, in terms of x,

   i $\cos k$ **ii** $\tan k$

Given, instead, that $-\frac{\pi}{2} < k < 0$,

c how, if at all, would it affect your answers to **b**?

7 The function f is defined as $f : x \rightarrow \arcsin x$, $-1 \leqslant x \leqslant 1$, and the function g is such that $g(x) = f(2x)$.

a Sketch the graph of $y = f(x)$ and state the range of f.

b Sketch the graph of $y = g(x)$.

c Define g in the form $g : x \rightarrow \ldots$ and give the domain of g.

d Define g^{-1} in the form $g^{-1} : x \rightarrow \ldots$

8 a Sketch the graph of $y = \sec x$, with the restricted domain $0 \leqslant x \leqslant \pi$, $x \neq \dfrac{\pi}{2}$.

b Given that $\mathrm{arcsec}\, x$ is the inverse function of $\sec x$, $0 \leqslant x \leqslant \pi$, $x \neq \dfrac{\pi}{2}$, sketch the graph of $y = \mathrm{arcsec}\, x$ and state the range of $\mathrm{arcsec}\, x$.

Mixed exercise 6F

Give any non-exact answers to equations to 1 decimal place.

1 Solve $\tan x = 2 \cot x$, in the interval $-180° \leqslant x \leqslant 90°$.

2 Given that $p = 2 \sec \theta$ and $q = 4 \cos \theta$, express p in terms of q.

3 Given that $p = \sin \theta$ and $q = 4 \cot \theta$, show that $p^2 q^2 = 16(1 - p^2)$.

4 a Solve, in the interval $0 < \theta < 180°$,

  i $\mathrm{cosec}\, \theta = 2 \cot \theta$ **ii** $2 \cot^2 \theta = 7 \mathrm{cosec}\, \theta - 8$

  b Solve, in the interval $0 \leqslant \theta \leqslant 360°$,

  i $\sec(2\theta - 15°) = \mathrm{cosec}\, 135°$ **ii** $\sec^2 \theta + \tan \theta = 3$

  c Solve, in the interval $0 \leqslant x \leqslant 2\pi$,

  i $\mathrm{cosec}\left(x + \dfrac{\pi}{15}\right) = -\sqrt{2}$ **ii** $\sec^2 x = \frac{4}{3}$

5 Given that $5 \sin x \cos y + 4 \cos x \sin y = 0$, and that $\cot x = 2$, find the value of $\cot y$.

6 Show that:

a $(\tan \theta + \cot \theta)(\sin \theta + \cos \theta) \equiv \sec \theta + \mathrm{cosec}\, \theta$

b $\dfrac{\mathrm{cosec}\, x}{\mathrm{cosec}\, x - \sin x} \equiv \sec^2 x$

c $(1 - \sin x)(1 + \mathrm{cosec}\, x) \equiv \cos x \cot x$

d $\dfrac{\cot x}{\mathrm{cosec}\, x - 1} - \dfrac{\cos x}{1 + \sin x} \equiv 2 \tan x$

e $\dfrac{1}{\mathrm{cosec}\, \theta - 1} + \dfrac{1}{\mathrm{cosec}\, \theta + 1} \equiv 2 \sec \theta \tan \theta$

f $\dfrac{(\sec \theta - \tan \theta)(\sec \theta + \tan \theta)}{1 + \tan^2 \theta} \equiv \cos^2 \theta$

7 a Show that $\dfrac{\sin x}{1 + \cos x} + \dfrac{1 + \cos x}{\sin x} \equiv 2\,\mathrm{cosec}\,x$.

[handwritten] $\dfrac{\sin^2 + 1 + \cos^2 + 2\cos}{\sin + \cos\sin}$

b Hence solve, in the interval $-2\pi \leqslant x \leqslant 2\pi$, $\dfrac{\sin x}{1 + \cos x} + \dfrac{1 + \cos x}{\sin x} = -\dfrac{4}{\sqrt{3}}$.

8 Prove that $\dfrac{1 + \cos\theta}{1 - \cos\theta} \equiv (\mathrm{cosec}\,\theta + \cot\theta)^2$.

9 Given that $\sec A = -3$, where $\dfrac{\pi}{2} < A < \pi$,

a calculate the exact value of $\tan A$.

b Show that $\mathrm{cosec}\,A = \dfrac{3\sqrt{2}}{4}$.

10 Given that $\sec\theta = k$, $|k| \geqslant 1$, and that θ is obtuse, express in terms of k:

a $\cos\theta$ **b** $\tan^2\theta$ **c** $\cot\theta$ **d** $\mathrm{cosec}\,\theta$

11 Solve, in the interval $0 \leqslant x \leqslant 2\pi$, the equation $\sec\left(x + \dfrac{\pi}{4}\right) = 2$, giving your answers in terms of π.

12 Find, in terms of π, the value of $\arcsin(\frac{1}{2}) - \arcsin(-\frac{1}{2})$.   *[handwritten]* $\dfrac{\pi}{3}$

13 Solve, in the interval $0 \leqslant x \leqslant 2\pi$, the equation $\sec^2 x - \dfrac{2\sqrt{3}}{3}\tan x - 2 = 0$, giving your answers in terms of π.

14 a Factorise $\sec x\,\mathrm{cosec}\,x - 2\sec x - \mathrm{cosec}\,x + 2$.

b Hence solve $\sec x\,\mathrm{cosec}\,x - 2\sec x - \mathrm{cosec}\,x + 2 = 0$, in the interval $0 \leqslant x \leqslant 360°$.

15 Given that $\arctan(x - 2) = -\dfrac{\pi}{3}$, find the value of x.   *[handwritten]* $x - 2 = \tan\left(\frac{-\pi}{3}\right)$ $x = \tan\left(\frac{-\pi}{3}\right) + 2$

16 On the same set of axes sketch the graphs of $y = \cos x$, $0 \leqslant x \leqslant \pi$, and $y = \arccos x$, $-1 \leqslant x \leqslant 1$, showing the coordinates of points in which the curves meet the axes.

17 a Given that $\sec x + \tan x = -3$, use the identity $1 + \tan^2 x \equiv \sec^2 x$ to find the value of $\sec x - \tan x$.

[handwritten] $\sec^2 - \tan^2 = 1$ $(\sec + \tan)(\sec - \tan) = 1$ $\sec - \tan = -\frac{1}{3}$

b Deduce the value of

i $\sec x$ **ii** $\tan x$

[handwritten] $-3 - \tan - \tan = -\frac{1}{3}$ $-2\tan = 2\frac{2}{3}$ $\tan = -1\frac{1}{3}$ $\sec = -\frac{5}{3}$

c Hence solve, in the interval $-180° \leqslant x \leqslant 180°$, $\sec x + \tan x = -3$.

[handwritten] $-53.1, 126.9, 26.9, 143.1$

18 Given that $p = \sec\theta - \tan\theta$ and $q = \sec\theta + \tan\theta$, show that $p = \dfrac{1}{q}$.

[handwritten] $\sec^2 - \tan^2 = 1$ $\tan^2 + 1 = \sec^2$ $(\sec - \tan)/(\sec + \tan)$

19 a Prove that $\sec^4\theta - \tan^4\theta \equiv \sec^2\theta + \tan^2\theta$.

b Hence solve, in the interval $-180° \leqslant \theta \leqslant 180°$, $\sec^4\theta = \tan^4\theta + 3\tan\theta$.

20 (Although integration is not in the specification for C3, this question only requires you to know that the area under a curve can be represented by an integral.)

a Sketch the graph of $y = \sin x$ and shade in the area representing $\displaystyle\int_0^{\frac{\pi}{2}} \sin x\,\mathrm{d}x$.

b Sketch the graph of $y = \arcsin x$ and shade in the area representing $\displaystyle\int_0^1 \arcsin x\,\mathrm{d}x$.

c By considering the shaded areas explain why $\displaystyle\int_0^{\frac{\pi}{2}} \sin x\,\mathrm{d}x + \int_0^1 \arcsin x\,\mathrm{d}x = \dfrac{\pi}{2}$.

Summary of key points

1 • $\sec\theta = \dfrac{1}{\cos\theta}$ {$\sec\theta$ is undefined when $\cos\theta = 0$, i.e. at $\theta = (2n + 1)\,90°$, $n \in \mathbb{Z}$}

 • $\operatorname{cosec}\theta = \dfrac{1}{\sin\theta}$ {$\operatorname{cosec}\theta$ is undefined when $\sin\theta = 0$, i.e. at $\theta = 180n°$, $n \in \mathbb{Z}$}

 • $\cot\theta = \dfrac{1}{\tan\theta}$ {$\cot\theta$ is undefined when $\tan\theta = 0$, i.e. at $\theta = 180n°$, $n \in \mathbb{Z}$}

 • $\cot\theta$ can also be written as $\dfrac{\cos\theta}{\sin\theta}$.

2 The graphs of $\sec\theta$, $\operatorname{cosec}\theta$ and $\cot\theta$ are

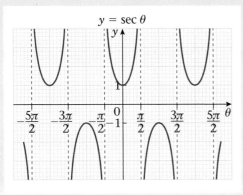

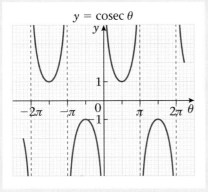

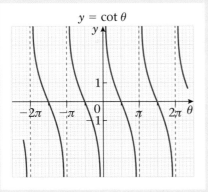

3 Two further Pythagorean identities, derived from $\sin^2\theta + \cos^2\theta \equiv 1$, are

$$1 + \tan^2\theta \equiv \sec^2\theta \quad \text{and} \quad 1 + \cot^2\theta \equiv \operatorname{cosec}^2\theta$$

4 The inverse function of $\sin x$, $-\dfrac{\pi}{2} \leqslant x \leqslant \dfrac{\pi}{2}$, is called $\arcsin x$;

it has domain $-1 \leqslant x \leqslant 1$ and range $-\dfrac{\pi}{2} \leqslant \arcsin x \leqslant \dfrac{\pi}{2}$

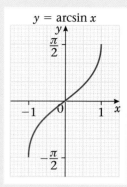

$y = \arcsin x$

5 The inverse function of $\cos x$, $0 \leqslant x \leqslant \pi$, is called $\arccos x$; it has domain $-1 \leqslant x \leqslant 1$ and range $0 \leqslant \arccos x \leqslant \pi$.

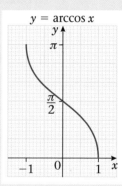

$y = \arccos x$

6 The inverse function of $\tan x$, $-\dfrac{\pi}{2} < x < \dfrac{\pi}{2}$, is called $\arctan x$;

it has domain $x \in \mathbb{R}$ and range $-\dfrac{\pi}{2} < \arctan x < \dfrac{\pi}{2}$.

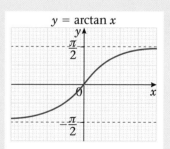

$y = \arctan x$

After completing this chapter you should be able to

1 use the addition formulae

2 use the double angle formulae

3 write expressions of the form $a\cos\theta \pm b\sin\theta$ in the form $R\cos(\theta \pm \alpha)$ and/or $R\sin(\theta \pm \alpha)$

4 use the factor formulae

5 use all of the above to solve equations and prove identities.

Further trigonometric identities and their applications

It is a common misconception amongst students that you treat $\sin(\theta + \alpha)$ in exactly the same way as $2(x + 3)$.

Whereas $2(x + 3) = 2x + 6$

$\sin(\theta + \alpha) \neq \sin\theta + \sin\alpha$

Whilst it is difficult to prove identities it is generally easier to prove they are false. All you need to do is find a counter example, a particular example that doesn't work.

7.1 You need to know and be able to use the addition formulae.

- $\sin(A + B) \equiv \sin A \cos B + \cos A \sin B$ $\sin(A - B) \equiv \sin A \cos B - \cos A \sin B$

- $\cos(A + B) \equiv \cos A \cos B - \sin A \sin B$ $\cos(A - B) \equiv \cos A \cos B + \sin A \sin B$

- $\tan(A + B) \equiv \dfrac{\tan A + \tan B}{1 - \tan A \tan B}$ $\tan(A - B) \equiv \dfrac{\tan A - \tan B}{1 + \tan A \tan B}$

Although you will not be expected to derive these formulae from first principles, it will help your understanding of them to see how one of them can be derived.

Example 1

Show that $\cos(A - B) \equiv \cos A \cos B + \sin A \sin B$

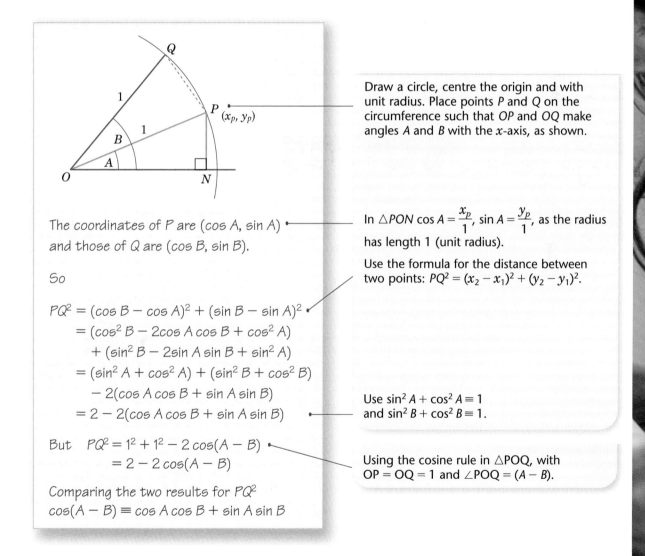

Draw a circle, centre the origin and with unit radius. Place points P and Q on the circumference such that OP and OQ make angles A and B with the x-axis, as shown.

The coordinates of P are $(\cos A, \sin A)$ and those of Q are $(\cos B, \sin B)$.

In $\triangle PON$ $\cos A = \dfrac{x_p}{1}$, $\sin A = \dfrac{y_p}{1}$, as the radius has length 1 (unit radius).

Use the formula for the distance between two points: $PQ^2 = (x_2 - x_1)^2 + (y_2 - y_1)^2$.

So

$$PQ^2 = (\cos B - \cos A)^2 + (\sin B - \sin A)^2$$
$$= (\cos^2 B - 2\cos A \cos B + \cos^2 A)$$
$$\quad + (\sin^2 B - 2\sin A \sin B + \sin^2 A)$$
$$= (\sin^2 A + \cos^2 A) + (\sin^2 B + \cos^2 B)$$
$$\quad - 2(\cos A \cos B + \sin A \sin B)$$
$$= 2 - 2(\cos A \cos B + \sin A \sin B)$$

Use $\sin^2 A + \cos^2 A \equiv 1$ and $\sin^2 B + \cos^2 B \equiv 1$.

But $PQ^2 = 1^2 + 1^2 - 2\cos(A - B)$
$$= 2 - 2\cos(A - B)$$

Using the cosine rule in $\triangle POQ$, with $OP = OQ = 1$ and $\angle POQ = (A - B)$.

Comparing the two results for PQ^2
$$\cos(A - B) \equiv \cos A \cos B + \sin A \sin B$$

The other formulae involving sine and cosine can be constructed using the one in the example above.

Example 2

Use the result in Example 1 to show that:

a $\cos(A + B) \equiv \cos A \cos B - \sin A \sin B$

b $\sin(A + B) \equiv \sin A \cos B + \cos A \sin B$

c $\sin(A - B) \equiv \sin A \cos B - \cos A \sin B$

a Replace B by $(-B)$ in
$$\cos(A - B) \equiv \cos A \cos B + \sin A \sin B$$
So $\cos(A + B) \equiv \cos A \cos(-B)$
$$+ \sin A \sin(-B)$$
∴ $\cos(A + B) \equiv \cos A \cos B - \sin A \sin B$

Use the results $\cos(-B) = \cos B$
and $\sin(-B) = -\sin B$
(See Chapter 8 in Book C2.)

b Replace A by $\left(\dfrac{\pi}{2} - A\right)$ in
$$\cos(A - B) \equiv \cos A \cos B + \sin A \sin B$$
So $\cos\left[\left(\dfrac{\pi}{2} - A\right) - B\right] \equiv \cos\left(\dfrac{\pi}{2} - A\right)\cos B$
$$+ \sin\left(\dfrac{\pi}{2} - A\right)\sin B$$
$$\cos\left[\dfrac{\pi}{2} - (A + B)\right] \equiv \cos\left(\dfrac{\pi}{2} - A\right)\cos B$$
$$+ \sin\left(\dfrac{\pi}{2} - A\right)\sin B$$
∴ $\sin(A + B) \equiv \sin A \cos B$
$$+ \cos A \sin B$$

Use $\cos\left(\dfrac{\pi}{2} - \theta\right) = \sin \theta$
and $\sin\left(\dfrac{\pi}{2} - \theta\right) = \cos \theta$
(See Chapter 8 in Book C2.)

c Replace B by $(-B)$ in the result in **b**:
$$\sin(A - B) \equiv \sin A \cos B + \cos A \sin(-B)$$
so $\sin(A - B) \equiv \sin A \cos B - \cos A \sin B$

To find similar expressions for $\tan(A + B)$ and $\tan(A - B)$ you can use the fact that $\tan \theta = \dfrac{\sin \theta}{\cos \theta}$ and divide the appropriate results given above.

Example 3

Show that

a $\tan(A + B) \equiv \dfrac{\tan A + \tan B}{1 - \tan A \tan B}$

b $\tan(A - B) \equiv \dfrac{\tan A - \tan B}{1 + \tan A \tan B}$

a $\tan(A + B) \equiv \dfrac{\sin(A + B)}{\cos(A + B)}$

$\equiv \dfrac{\sin A \cos B + \cos A \sin B}{\cos A \cos B - \sin A \sin B}$

Dividing the 'top and bottom' by $\cos A \cos B$ gives

$\tan(A + B)$

$\equiv \dfrac{\dfrac{\sin A \cancel{\cos B}}{\cos A \cancel{\cos B}} + \dfrac{\cancel{\cos A} \sin B}{\cancel{\cos A} \cos B}}{\dfrac{\cancel{\cos A} \cancel{\cos B}}{\cancel{\cos A} \cancel{\cos B}} - \dfrac{\sin A \sin B}{\cos A \cos B}}$

Cancel terms, as shown, and use the result $\tan \theta = \dfrac{\sin \theta}{\cos \theta}$

$\equiv \dfrac{\tan A + \tan B}{1 - \tan A \tan B}$

b Replace B by $-B$ in the result above:

$\tan(A - B) \equiv \dfrac{\tan A + \tan(-B)}{1 - \tan A \tan(-B)}$

Use the result $\tan(-\theta) = -\tan \theta$.
See Chapter 8 in Book C2.

$\equiv \dfrac{\tan A - \tan B}{1 + \tan A \tan B}$

Example 4

Show, using the formula for $\sin(A - B)$, that $\sin 15° = \dfrac{\sqrt{6} - \sqrt{2}}{4}$

$\sin 15° = \sin(45 - 30)°$

$= \sin 45° \cos 30° - \cos 45° \sin 30°$

$= (\tfrac{1}{2}\sqrt{2})(\tfrac{1}{2}\sqrt{3}) - (\tfrac{1}{2}\sqrt{2})(\tfrac{1}{2})$

$= \tfrac{1}{4}(\sqrt{3}\sqrt{2} - \sqrt{2})$

$= \tfrac{1}{4}(\sqrt{6} - \sqrt{2})$

You know the exact form of sin and cos for many angles, e.g. 30°, 45°, 60°, 90°, 180°…, so write 15° using two of these angles.
[You could equally use $\sin(60 - 45)°$.]

Example 5

Given that $\sin A = -\tfrac{3}{5}$ and $180° < A < 270°$, and that $\cos B = -\tfrac{12}{13}$ and B is obtuse, find the value of

a $\cos(A - B)$

b $\tan(A + B)$

a $\cos(A - B) \equiv \cos A \cos B + \sin A \sin B$

$\cos^2 A \equiv 1 - \sin^2 A = 1 - \left(-\frac{3}{5}\right)^2 = \frac{16}{25}$

So $\cos A = \pm\frac{4}{5}$

You need to find $\cos A$ and $\sin B$.
Take note of the quadrants that A and B are in.

You can also work with the associated acute angles A' and B'. Draw right-angled triangles and use Pythagoras' theorem.

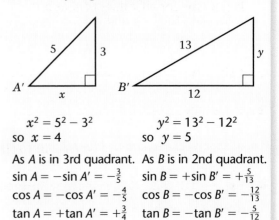

$x^2 = 5^2 - 3^2$ $y^2 = 13^2 - 12^2$
so $x = 4$ so $y = 5$

but A is in the third quadrant, where $\cos$ is $-$ve,

$\therefore \quad \cos A = -\frac{4}{5}$

$\sin^2 B \equiv 1 - \cos^2 B = 1 - \left(-\frac{12}{13}\right)^2 = \frac{25}{169}$

So $\sin B = \pm\frac{5}{13}$

but B is in the 2nd quadrant

$\therefore \quad \sin B = +\frac{5}{13}$

$\cos(A - B) = \left(-\frac{4}{5}\right)\left(-\frac{12}{13}\right) + \left(-\frac{3}{5}\right)\left(+\frac{5}{13}\right)$

$= \frac{33}{65}$

As A is in 3rd quadrant.   As B is in 2nd quadrant.

$\sin A = -\sin A' = -\frac{3}{5}$ $\sin B = +\sin B' = +\frac{5}{13}$

$\cos A = -\cos A' = -\frac{4}{5}$ $\cos B = -\cos B' = -\frac{12}{13}$

$\tan A = +\tan A' = +\frac{3}{4}$ $\tan B = -\tan B' = -\frac{5}{12}$

b $\tan(A + B) \equiv \dfrac{\tan A + \tan B}{1 - \tan A \tan B}$

So $\tan(A + B) = \dfrac{\frac{3}{4} + \left(-\frac{5}{12}\right)}{1 - \left(\frac{3}{4}\right)\left(-\frac{5}{12}\right)}$

$= \dfrac{\frac{1}{3}}{\frac{63}{48}} = \frac{1}{3} \times \frac{48}{63} = \frac{16}{63}$

You can use the above results for $\tan A$ and $\tan B$, or use $\tan \theta = \dfrac{\sin \theta}{\cos \theta}$ with the results for $\sin A$, $\cos A$, $\sin B$ and $\cos B$.

Example 6

Given that $2 \sin(x + y) = 3 \cos(x - y)$, express $\tan x$ in terms of $\tan y$.

Expanding $\sin(x + y)$ and $\cos(x - y)$ gives

$2 \sin x \cos y + 2 \cos x \sin y$

$= 3 \cos x \cos y + 3 \sin x \sin y$

so $\dfrac{2 \sin x \cos y}{\cos x \cos y} + \dfrac{2 \cos x \sin y}{\cos x \cos y}$

$= \dfrac{3 \cos x \cos y}{\cos x \cos y} + \dfrac{3 \sin x \sin y}{\cos x \cos y}$

$2 \tan x + 2 \tan y = 3 + 3 \tan x \tan y$

$2 \tan x - 3 \tan x \tan y = 3 - 2 \tan y$

$\tan x(2 - 3 \tan y) = 3 - 2 \tan y$

So $\qquad\qquad \tan x = \dfrac{3 - 2 \tan y}{2 - 3 \tan y}$

This is similar to the expression seen in deriving $\tan(A + B)$. A good strategy is to divide both sides by $\cos x \cos y$.

Collect all terms in $\tan x$ on one side.

Factorise.

Exercise 7A

1 A student makes the mistake of thinking that
$\sin(A + B) \equiv \sin A + \sin B$.
Choose non-zero values of A and B to show that this statement is not true for all values of A and B.

> This is a very common error – don't make the same mistake. One counterexample is sufficient to disprove a statement.

2 Using the expansion of $\cos(A - B)$ with $A = B = \theta$, show that $\sin^2 \theta + \cos^2 \theta \equiv 1$.

$\cos(A - B) = \cos A \cos A + \sin A \sin A = 1$

3 a Use the expansion of $\sin(A - B)$ to show that $\sin\left(\dfrac{\pi}{2} - \theta\right) = \cos \theta$.

b Use the expansion of $\cos(A - B)$ to show that $\cos\left(\dfrac{\pi}{2} - \theta\right) = \sin \theta$.

4 Express the following as a single sine, cosine or tangent:

a $\sin 15° \cos 20° + \cos 15° \sin 20°$

b $\sin 58° \cos 23° - \cos 58° \sin 23°$

c $\cos 130° \cos 80° - \sin 130° \sin 80°$

d $\dfrac{\tan 76° - \tan 45°}{1 + \tan 76° \tan 45°}$

e $\cos 2\theta \cos \theta + \sin 2\theta \sin \theta$ $\cos(2\theta - \theta)$

f $\cos 4\theta \cos 3\theta - \sin 4\theta \sin 3\theta$ $\cos(7\theta)$

g $\sin \tfrac{1}{2}\theta \cos 2\tfrac{1}{2}\theta + \cos \tfrac{1}{2}\theta \sin 2\tfrac{1}{2}\theta$ $\sin(\theta)$

h $\dfrac{\tan 2\theta + \tan 3\theta}{1 - \tan 2\theta \tan 3\theta}$

i $\sin(A + B) \cos B - \cos(A + B) \sin B$

j $\cos\left(\dfrac{3x + 2y}{2}\right)\cos\left(\dfrac{3x - 2y}{2}\right) - \sin\left(\dfrac{3x + 2y}{2}\right)\sin\left(\dfrac{3x - 2y}{2}\right)$

5 Calculate, without using your calculator, the exact value of:

a $\sin 30° \cos 60° + \cos 30° \sin 60°$ $\sin(90)$

b $\cos 110° \cos 20° + \sin 110° \sin 20°$ $\cos(90)$

c $\sin 33° \cos 27° + \cos 33° \sin 27°$

d $\cos\dfrac{\pi}{8}\cos\dfrac{\pi}{8} - \sin\dfrac{\pi}{8}\sin\dfrac{\pi}{8}$ $\cos(45)$

e $\sin 60° \cos 15° - \cos 60° \sin 15°$

f $\cos 70°(\cos 50° - \tan 70° \sin 50°)$

g $\dfrac{\tan 45° + \tan 15°}{1 - \tan 45° \tan 15°}$

h $\dfrac{1 - \tan 15°}{1 + \tan 15°}$ $\tan(30)$ **Hint:** $\tan 45° = 1$.

i $\dfrac{\tan\left(\dfrac{7\pi}{12}\right) - \tan\left(\dfrac{\pi}{3}\right)}{1 + \tan\left(\dfrac{7\pi}{12}\right)\tan\left(\dfrac{\pi}{3}\right)}$

j $\sqrt{3}\cos 15° - \sin 15°$ **Hint:** Look at **e**.

6 Triangle ABC is such that $AB = 3$ cm, $BC = 4$ cm, $\angle ABC = 120°$ and $\angle BAC = \theta°$.

a Write down, in terms of θ, an expression for $\angle ACB$.

b Using the sine rule, or otherwise, show that $\tan \theta° = \dfrac{2\sqrt{3}}{5}$.

7 Prove the identities

a $\sin(A + 60°) + \sin(A - 60°) \equiv \sin A$

b $\dfrac{\cos A}{\sin B} - \dfrac{\sin A}{\cos B} \equiv \dfrac{\cos(A + B)}{\sin B \cos B}$

c $\dfrac{\sin(x + y)}{\cos x \cos y} \equiv \tan x + \tan y$

d $\dfrac{\cos(x + y)}{\sin x \sin y} + 1 \equiv \cot x \cot y$

e $\cos\left(\theta + \dfrac{\pi}{3}\right) + \sqrt{3}\sin\theta \equiv \sin\left(\theta + \dfrac{\pi}{6}\right)$

f $\cot(A + B) \equiv \dfrac{\cot A \cot B - 1}{\cot A + \cot B}$

g $\sin^2(45 + \theta)° + \sin^2(45 - \theta)° \equiv 1$

h $\cos(A + B)\cos(A - B) \equiv \cos^2 A - \sin^2 B$

8 Given that $\sin A = \frac{4}{5}$ and $\sin B = \frac{1}{2}$, where A and B are both acute angles, calculate the exact values of

a $\sin(A + B)$ **b** $\cos(A - B)$ **c** $\sec(A - B)$

9 Given that $\cos A = -\frac{4}{5}$, and A is an obtuse angle measured in radians, find the exact value of

a $\sin A$ **b** $\cos(\pi + A)$ **c** $\sin\left(\dfrac{\pi}{3} + A\right)$ **d** $\tan\left(\dfrac{\pi}{4} + A\right)$

10 Given that $\sin A = \frac{8}{17}$, where A is acute, and $\cos B = -\frac{4}{5}$, where B is obtuse, calculate the exact value of

a $\sin(A - B)$ **b** $\cos(A - B)$ **c** $\cot(A - B)$

11 Given that $\tan A = \frac{7}{24}$, where A is reflex, and $\sin B = \frac{5}{13}$, where B is obtuse, calculate the exact value of

a $\sin(A + B)$ **b** $\tan(A - B)$ **c** $\mathrm{cosec}(A + B)$

12 Write the following as a single trigonometric function, assuming that θ is measured in radians:

a $\cos^2\theta - \sin^2\theta$ **b** $2\sin 4\theta\cos 4\theta$ **c** $\dfrac{1 + \tan\theta}{1 - \tan\theta}$ **d** $\dfrac{1}{\sqrt{2}}(\sin\theta + \cos\theta)$

13 Solve, in the interval $0° \leqslant \theta < 360°$, the following equations. Give answers to the nearest $0.1°$.

a $3\cos\theta = 2\sin(\theta + 60°)$
c $\cos(\theta + 25°) + \sin(\theta + 65°) = 1$
e $\tan(\theta - 45°) = 6\tan\theta$

b $\sin(\theta + 30°) + 2\sin\theta = 0$
d $\cos\theta = \cos(\theta + 60°)$
f $\sin\theta + \cos\theta = 1$

Hint for part **f**: Multiply each term by $\dfrac{1}{\sqrt{2}}$

14 a Solve the equation $\cos\theta\cos 30° - \sin\theta\sin 30° = 0.5$, for $0 \leqslant \theta \leqslant 360°$.
b Hence write down, in the same interval, the solutions of $\sqrt{3}\cos\theta - \sin\theta = 1$.

15 a Express $\tan(45 + 30)°$ in terms of $\tan 45°$ and $\tan 30°$.
b Hence show that $\tan 75° = 2 + \sqrt{3}$.

16 Show that $\sec 105° = -\sqrt{2}(1 + \sqrt{3})$

17 Calculate the exact values of

a $\cos 15°$ **b** $\sin 75°$ **c** $\sin(120 + 45)°$ **d** $\tan 165°$

Hint for part **a**: Write $15°$ as $(45 - 30)°$

18 a Given that $3\sin(x - y) - \sin(x + y) = 0$, show that $\tan x = 2\tan y$.
b Solve $3\sin(x - 45°) - \sin(x + 45°) = 0$, for $0 \leqslant x \leqslant 360°$.

19 Given that $\sin x(\cos y + 2\sin y) = \cos x(2\cos y - \sin y)$, find the value of $\tan(x + y)$.

20 Given that $\tan(x - y) = 3$, express $\tan y$ in terms of $\tan x$.

21 In each of the following, calculate the exact value of $\tan x°$.
 a $\tan(x - 45)° = \frac{1}{4}$
 b $\sin(x - 60)° = 3\cos(x + 30)°$
 c $\tan(x - 60)° = 2$

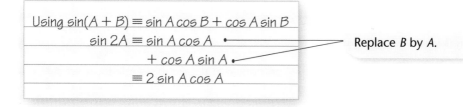

$$\frac{\tan x - 1}{1 + \tan x} = \frac{1}{4} + \frac{1}{4}\tan x$$
$$\tan x = \frac{3}{4}$$

22 Given that $\tan A° = \frac{1}{5}$ and $\tan B° = \frac{2}{3}$, calculate, without using your calculator, the value of $A + B$,
 a where A and B are both acute,
 b where A is reflex and B is acute.

23 Given that $\cos y = \sin(x + y)$, show that $\tan y = \sec x - \tan x$.

24 Given that $\cot A = \frac{1}{4}$ and $\cot(A + B) = 2$, find the value of $\cot B$.

25 Given that $\tan\left(x + \frac{\pi}{3}\right) = \frac{1}{2}$, show that $\tan x = 8 - 5\sqrt{3}$.

7.2 **You can express $\sin 2A$, $\cos 2A$ and $\tan 2A$ in terms of angle A, using the double angle formulae.**

■ $\sin 2A \equiv 2\sin A\cos A$
■ $\cos 2A \equiv \cos^2 A - \sin^2 A \equiv 2\cos^2 A - 1 \equiv 1 - 2\sin^2 A$
■ $\tan 2A \equiv \dfrac{2\tan A}{1 - \tan^2 A}$

You can derive these results from the addition formulae.

Example 7

Use the expansion of $\sin(A + B)$ to show that $\sin 2A \equiv 2\sin A\cos A$

Using $\sin(A + B) \equiv \sin A\cos B + \cos A\sin B$
$\sin 2A \equiv \sin A\cos A$ Replace B by A.
       $+ \cos A\sin A$
      $\equiv 2\sin A\cos A$

Example 8

 a By using an appropriate addition formula show that $\cos 2A \equiv \cos^2 A - \sin^2 A$.
 b Hence derive the alternative forms $\cos 2A \equiv 2\cos^2 A - 1 \equiv 1 - 2\sin^2 A$.

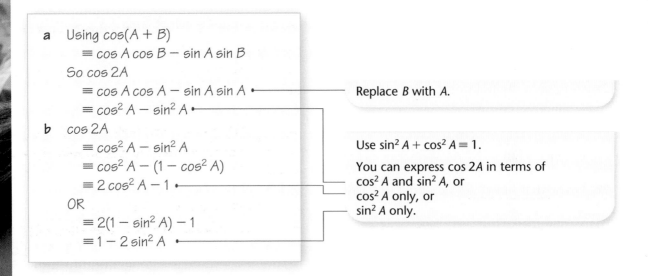

a Using $\cos(A + B)$
$$\equiv \cos A \cos B - \sin A \sin B$$
So $\cos 2A$
$$\equiv \cos A \cos A - \sin A \sin A$$ ———— Replace B with A.
$$\equiv \cos^2 A - \sin^2 A$$

b $\cos 2A$
$$\equiv \cos^2 A - \sin^2 A$$
$$\equiv \cos^2 A - (1 - \cos^2 A)$$ Use $\sin^2 A + \cos^2 A \equiv 1$.
$$\equiv 2\cos^2 A - 1$$ You can express $\cos 2A$ in terms of $\cos^2 A$ and $\sin^2 A$, or $\cos^2 A$ only, or $\sin^2 A$ only.
OR
$$\equiv 2(1 - \sin^2 A) - 1$$
$$\equiv 1 - 2\sin^2 A$$

Example 9

Express $\tan 2A$ in terms of $\tan A$.

Using $\tan(A + B) \equiv \dfrac{\tan A + \tan B}{1 - \tan A \tan B}$

$$\tan 2A \equiv \frac{2\tan A}{1 - \tan^2 A}$$ ———— Replace B by A.

Example 10

Rewrite the following expressions as a single trigonometric function:

a $2\sin\dfrac{\theta}{2}\cos\dfrac{\theta}{2}\cos\theta$

b $1 + \cos 4\theta$

Using $2\sin A \cos A \equiv \sin 2A$ with $A = \dfrac{\theta}{2}$.

a $2\sin\dfrac{\theta}{2}\cos\dfrac{\theta}{2} \equiv \sin\theta$

So $2\sin\dfrac{\theta}{2}\cos\dfrac{\theta}{2}\cos\theta \equiv \sin\theta\cos\theta$

$$\equiv \tfrac{1}{2}\sin 2\theta$$

The double angle formulae allow you to convert an angle into its half angle, and vice versa. In a question it will not always be obvious that the double angle formulae are needed (i.e. you will not always see **2A** or **2θ**).

$\sin 2\theta = 2\sin\theta\cos\theta$.

b $\cos 4\theta \equiv 2\cos^2 2\theta - 1$

So $1 + \cos 4\theta \equiv 2\cos^2 2\theta$

Using $\cos 2A = 2\cos^2 A - 1$ with $A = 2\theta$. Choose this form of $\cos 2A$ because the -1 will cancel out the $+1$ in '$1 + \cos 4\theta$'.

Example 11

Given that $\cos x = \frac{3}{4}$, and that $180° < x < 360°$, find the exact values of

a $\sin 2x$ **b** $\tan 2x$

a

Draw the right-angled triangle with $\cos x' = \frac{3}{4}$.

Using Pythagoras' theorem

$$y^2 = 4^2 - 3^2 = 7$$

So $y = \sqrt{7}$

$\therefore \quad \sin x' = \dfrac{\sqrt{7}}{4}$ and $\tan x' = \dfrac{\sqrt{7}}{3}$

$\Rightarrow \quad \sin x = -\dfrac{\sqrt{7}}{4}$ and $\tan x = -\dfrac{\sqrt{7}}{3}$

As $\cos x$ is +ve and x is reflex, x must be in the 4th quadrant, so $\sin x = -\sin x'$ and $\tan x = -\tan x'$.

$\sin 2x = 2 \sin x \cos x$

$$= 2\left(-\frac{\sqrt{7}}{4}\right)\left(\frac{3}{4}\right) = -\frac{3\sqrt{7}}{8}$$

b $\tan 2x = \dfrac{2 \tan x}{1 - \tan^2 x} = \dfrac{-\dfrac{2\sqrt{7}}{3}}{1 - \dfrac{7}{9}}$

$$= -\frac{2\sqrt{7}}{3} \times \frac{9}{2}$$

$$= -3\sqrt{7}$$

Exercise 7B

In equations, give answers to 1 decimal place where appropriate.

1 Write the following expressions as a single trigonometric ratio:

a $2 \sin 10° \cos 10°$ $= \sin 5$

b $1 - 2 \sin^2 25°$ $\cos 50$

c $\cos^2 40° - \sin^2 40°$

d $\dfrac{2 \tan 5°}{1 - \tan^2 5°}$

e $\dfrac{1}{2 \sin(24\frac{1}{2})° \cos(24\frac{1}{2})°}$ $= \dfrac{1}{\sin(49)}$

f $6 \cos^2 30° - 3$

g $\dfrac{\sin 8°}{\sec 8°}$ $\dfrac{1}{2} \sin 16$

h $\cos^2 \dfrac{\pi}{16} - \sin^2 \dfrac{\pi}{16}$

2 Without using your calculator find the exact values of:

a $2\sin(22\tfrac{1}{2})° \cos(22\tfrac{1}{2})°$ **b** $2\cos^2 15° - 1$ **c** $(\sin 75° - \cos 75°)^2$ **d** $\dfrac{2\tan\dfrac{\pi}{8}}{1 - \tan^2\dfrac{\pi}{8}}$

Sin 45

tan 1 tan 45

3 Write the following in their simplest form, involving only one trigonometric function:

a $\cos^2 3\theta - \sin^2 3\theta$

b $6\sin 2\theta \cos 2\theta$

3 sin 4θ

c $\dfrac{2\tan\dfrac{\theta}{2}}{1 - \tan^2\dfrac{\theta}{2}}$

d $2 - 4\sin^2\dfrac{\theta}{2}$

e $\sqrt{1 + \cos 2\theta}$ *√2cos²θ*

f $\sin^2\theta \cos^2\theta$

g $4\sin\theta \cos\theta \cos 2\theta$

h $\dfrac{\tan\theta}{\sec^2\theta - 2}$ *$\dfrac{\tan\theta}{\frac{1}{\cos^2\theta} - 2}$*

i $\sin^4\theta - 2\sin^2\theta \cos^2\theta + \cos^4\theta$

4 Given that $\cos x = \frac{1}{4}$, find the exact value of $\cos 2x$.

5 Find the possible values of $\sin\theta$ when $\cos 2\theta = \frac{23}{25}$. *$1 - 2\sin^2\theta = \frac{23}{25}$*

$\pm\frac{1}{5} = \sin\theta$

6 Given that $\cos x + \sin x = m$ and $\cos x - \sin x = n$, where m and n are constants, write down, in terms of m and n, the value of $\cos 2x$.

$2 \times \frac{3}{5} \times \frac{4}{5} = \frac{24}{25}$ *$\frac{16}{25} - \frac{9}{25} = \frac{7}{25}$*   *$\frac{5}{3}$*

7 Given that $\tan\theta = \frac{3}{4}$, and that θ is acute:

a Find the exact value of **i** $\tan 2\theta$ **ii** $\sin 2\theta$ **iii** $\cos 2\theta$

4

b Deduce the value of $\sin 4\theta$. *$2 \times \frac{24}{25} \times \frac{7}{25} = \frac{336}{625}$*

8 Given that $\cos A = -\frac{1}{3}$, and that A is obtuse:

a Find the exact value of **i** $\cos 2A$ **ii** $\sin A$ **iii** $\operatorname{cosec} 2A$

b Show that $\tan 2A = \dfrac{4\sqrt{2}}{7}$.

9 Given that $\pi < \theta < \dfrac{3\pi}{2}$, find the value of $\tan\dfrac{\theta}{2}$ when $\tan\theta = \frac{3}{4}$.

10 In $\triangle ABC$, $AB = 4$ cm, $AC = 5$ cm, $\angle ABC = 2\theta$ and $\angle ACB = \theta$. Find the value of θ, giving your answer, in degrees, to 1 decimal place.

11 In $\triangle PQR$, $PQ = 3$ cm, $PR = 6$ cm, $QR = 5$ cm and $\angle QPR = 2\theta$.

a Use the cosine rule to show that $\cos 2\theta = \frac{5}{9}$.

b Hence find the exact value of $\sin\theta$.

12 The line l, with equation $y = \frac{3}{4}x$, bisects the angle between the x-axis and the line $y = mx$, $m > 0$. Given that the scales on each axis are the same, and that l makes an angle θ with the x-axis,

a write down the value of $\tan\theta$.

b Show that $m = \frac{24}{7}$.

7.3 The double angle formulae allow you to solve more equations and prove more identities.

Example 12

Prove the identity $\tan 2\theta \equiv \dfrac{2}{\cot \theta - \tan \theta}$

Start on LHS with $\tan 2\theta \equiv \dfrac{2\tan\theta}{1 - \tan^2 \theta}$

Divide 'top and bottom' by $\tan \theta$.

So $\tan 2\theta \equiv \dfrac{2}{\dfrac{1}{\tan \theta} - \tan \theta}$

$\equiv \dfrac{2}{\cot \theta - \tan \theta}$

There are many starting points here; the more you know the more options you have.

Try starting with $\tan 2\theta \equiv \dfrac{\sin 2\theta}{\cos 2\theta}$ and use the double angle formulae for $\sin 2\theta$ and $\cos 2\theta$ (a bit harder!), or start with the RHS.

Example 13

By expanding $\sin(2A + A)$ show that $\sin 3A \equiv 3 \sin A - 4 \sin^3 A$.

$\sin(2A + A) \equiv \sin 2A \cos A + \cos 2A \sin A$
So $\sin 3A \equiv (2 \sin A \cos A) \cos A$
$\quad + (\cos^2 A - \sin^2 A) \sin A$
$\equiv 2 \sin A \cos^2 A$
$\quad + \cos^2 A \sin A - \sin^3 A$
$\equiv 3 \sin A \cos^2 A - \sin^3 A \quad *$
$\equiv 3 \sin A(1 - \sin^2 A) - \sin^3 A$
$\equiv 3 \sin A - 4 \sin^3 A$

$\cos 2A = \cos^2 A - \sin^2 A$ has been used here because the line marked * is a useful result when you need to find the formula for $\tan 3A$.

See Exercise 7C Questions 12 and 13 for similar expansions of $\cos 3A$ and $\tan 3A$.

Example 14

Given that $x = 3 \sin \theta$ and $y = 3 - 4 \cos 2\theta$, eliminate θ and express y in terms of x.

The equations can be rewritten as

$\sin \theta = \dfrac{x}{3} \quad \cos 2\theta = \dfrac{3 - y}{4}$

As $\cos 2\theta = 1 - 2 \sin^2 \theta$ for all values of θ,

$\dfrac{3 - y}{4} = 1 - 2\left(\dfrac{x}{3}\right)^2$

So $\dfrac{y}{4} = 2\left(\dfrac{x}{3}\right)^2 - \dfrac{1}{4}$

or $y = 8\left(\dfrac{x}{3}\right)^2 - 1$

Be careful with this manipulation. Many errors occur in the early part of a solution.

This is the relationship: θ has been eliminated but the solution is not complete. Always make sure that you have answered the question: here you need to write $y = \ldots$.

Example 15

Solve $3\cos 2x - \cos x + 2 = 0$ for $0° \leqslant x \leqslant 360°$.

Using a double angle formula for $\cos 2x$

$$3\cos 2x - \cos x + 2 = 0$$

becomes

$$3(2\cos^2 x - 1) - \cos x + 2 = 0$$
$$6\cos^2 x - 3 - \cos x + 2 = 0$$
$$6\cos^2 x - \cos x - 1 = 0$$
So $(3\cos x + 1)(2\cos x - 1) = 0$

Solving: $\cos x = -\frac{1}{3}$ or $\cos x = \frac{1}{2}$

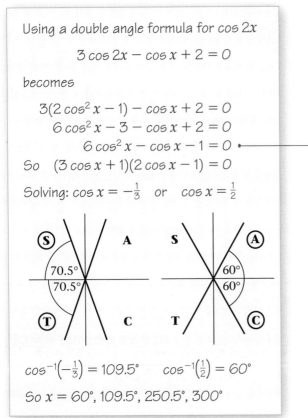

$$\cos^{-1}\left(-\tfrac{1}{3}\right) = 109.5° \qquad \cos^{-1}\left(\tfrac{1}{2}\right) = 60°$$

So $x = 60°, 109.5°, 250.5°, 300°$

The term $\cos x$ in the equation dictates the choice you need to make; you need the form of $\cos 2x$ with $\cos x$ only, i.e. $\cos 2x \equiv 2\cos^2 x - 1$.

This quadratic equation factorises $6X^2 - X - 1 = (3X + 1)(2X - 1)$.

Solutions to $\cos x = -\frac{1}{3}$ are in the 2nd and 3rd quadrants. Solutions to $\cos x = \frac{1}{2}$ are in the 1st and 4th quadrants.

Remember that two solutions of $\cos x = k$ are $\cos^{-1} k$ and $360° - \cos^{-1} k$. In this case they all fall in the required interval.

Exercise 7C

In equations, give answers to 1 decimal place where appropriate.

1 Prove the following identities:

a $\dfrac{\cos 2A}{\cos A + \sin A} \equiv \cos A - \sin A$ *Cos A − Sin A*

b $\dfrac{\sin B}{\sin A} - \dfrac{\cos B}{\cos A} \equiv 2\operatorname{cosec} 2A \sin(B - A)$

c $\dfrac{1 - \cos 2\theta}{\sin 2\theta} \equiv \tan \theta$

d $\dfrac{\sec^2 \theta}{1 - \tan^2 \theta} \equiv \sec 2\theta$

e $2(\sin^3 \theta \cos \theta + \cos^3 \theta \sin \theta) \equiv \sin 2\theta$
   (Sin Cos) (1) ≡ Sin 2θ

f $\dfrac{\sin 3\theta}{\sin \theta} - \dfrac{\cos 3\theta}{\cos \theta} \equiv 2$

g $\operatorname{cosec} \theta - 2\cot 2\theta \cos \theta \equiv 2\sin \theta$

h $\dfrac{\sec \theta - 1}{\sec \theta + 1} \equiv \tan^2 \dfrac{\theta}{2}$

i $\tan\left(\dfrac{\pi}{4} - x\right) \equiv \dfrac{1 - \sin 2x}{\cos 2x}$

2 a Show that $\tan \theta + \cot \theta \equiv 2\operatorname{cosec} 2\theta$.

    b Hence find the value of $\tan 75° + \cot 75°$.

3 Solve the following equations, in the interval shown in brackets:

    a $\sin 2\theta = \sin \theta$ $\{0 \le \theta \le 2\pi\}$

    b $\cos 2\theta = 1 - \cos \theta$ $\{-180° < \theta \le 180°\}$

    c $3 \cos 2\theta = 2 \cos^2 \theta$ $\{0 \le \theta < 360°\}$

    d $\sin 4\theta = \cos 2\theta$ $\{0 \le \theta \le \pi\}$

    e $2 \tan 2y \tan y = 3$ $\{0 \le y < 360°\}$

    f $3 \cos\theta - \sin \dfrac{\theta}{2} - 1 = 0$ $\{0 \le \theta < 720°\}$

    g $\cos^2 \theta - \sin 2\theta = \sin^2 \theta$ $\{0 \le \theta \le \pi\}$ $Cos2\theta = Sin2\theta$

    h $2 \sin\theta = \sec \theta$ $\{0 \le \theta \le 2\pi\}$ $Sin2\theta = 1$ $\tan 2\theta = 1$

        $\theta = 45,$ $\theta = \dfrac{\pi}{8}, \dfrac{5\pi}{8}$

    i $2 \sin 2\theta = 3 \tan\theta$ $\{0 \le \theta < 360°\}$

    j $2 \tan\theta = \sqrt{3}(1 - \tan \theta)(1 + \tan \theta)$ $\{0 \le \theta \le 2\pi\}$

    k $5 \sin 2\theta + 4 \sin\theta = 0$ $\{-180° < \theta \le 180°\}$ $10 Sin Cos + 4 Sin\theta$ $2 Sin\theta(5Cos + 2) = 0$

    l $\sin^2 \theta = 2 \sin 2\theta$ $\{-180° < \theta \le 180°\}$

    m $4 \tan\theta = \tan 2\theta$ $\{0 \le \theta < 360°\}$

4 Given that $p = 2 \cos\theta$ and $q = \cos 2\theta$, express q in terms of p.

5 Eliminate θ from the following pairs of equations:

    a $x = \cos^2 \theta,\ y = 1 - \cos 2\theta$ $y = x$ **b** $x = \tan \theta,\ y = \cot 2\theta$ $\dfrac{Cos2\theta}{Sin2\theta}$

    c $x = \sin \theta,\ y = \sin 2\theta$ **d** $x = 3 \cos 2\theta + 1,\ y = 2 \sin\theta$

              $2 \sin \cos$

6 **a** Prove that $(\cos 2\theta - \sin 2\theta)^2 \equiv 1 - \sin 4\theta$.

    b Use the result to solve, for $0 \le \theta < \pi$, the equation $\cos 2\theta - \sin 2\theta = \dfrac{1}{\sqrt{2}}$.
      Give your answers in terms of π.

7 **a** Show that:

    i $\sin \theta \equiv \dfrac{2 \tan \dfrac{\theta}{2}}{1 + \tan^2 \dfrac{\theta}{2}}$ **ii** $\cos \theta \equiv \dfrac{1 - \tan^2 \dfrac{\theta}{2}}{1 + \tan^2 \dfrac{\theta}{2}}$

    b By writing the following equations as quadratics in $\tan \dfrac{\theta}{2}$, solve, in the interval
    $0 \le \theta \le 360°$:

    i $\sin \theta + 2 \cos \theta = 1$ **ii** $3 \cos \theta - 4 \sin \theta = 2$

8 **a** Using $\cos 2A \equiv 2 \cos^2 A - 1 \equiv 1 - 2 \sin^2 A$, show that:

    i $\cos^2 \dfrac{x}{2} \equiv \dfrac{1 + \cos x}{2}$ **ii** $\sin^2 \dfrac{x}{2} \equiv \dfrac{1 - \cos x}{2}$

    b Given that $\cos \theta = 0.6$, and that θ is acute, write down the values of:

    i $\cos \dfrac{\theta}{2}$ **ii** $\sin \dfrac{\theta}{2}$ **iii** $\tan \dfrac{\theta}{2}$

    c Show that $\cos^4 \dfrac{A}{2} \equiv \tfrac{1}{8}(3 + 4 \cos A + \cos 2A)$

> These are known as the **half angle formulae**. (They are useful in integration.)

9 **a** Show that $3\cos^2 x - \sin^2 x \equiv 1 + 2\cos 2x$.

   b Hence sketch, for $-\pi \leqslant x \leqslant \pi$, the graph of $y = 3\cos^2 x - \sin^2 x$, showing the coordinates of points where the curve meets the axes.

10 **a** Express $2\cos^2 \dfrac{\theta}{2} - 4\sin^2 \dfrac{\theta}{2}$ in the form $a\cos\theta + b$, where a and b are constants.

   b Hence solve $2\cos^2 \dfrac{\theta}{2} - 4\sin^2 \dfrac{\theta}{2} = -3$, in the interval $0 \leqslant \theta < 360°$.

11 **a** Use the identity $\sin^2 A + \cos^2 A \equiv 1$ to show that $\sin^4 A + \cos^4 A \equiv \frac{1}{2}(2 - \sin^2 2A)$.

   b Deduce that $\sin^4 A + \cos^4 A \equiv \frac{1}{4}(3 + \cos 4A)$.

   c Hence solve $8\sin^4 \theta + 8\cos^4 \theta = 7$, for $0 < \theta < \pi$.

12 **a** By expanding $\cos(2A + A)$ show that $\cos 3A \equiv 4\cos^3 A - 3\cos A$.

   b Hence solve $8\cos^3 \theta - 6\cos\theta - 1 = 0$, for $\{0 \leqslant \theta \leqslant 360°\}$.

13 **a** Show that $\tan 3\theta \equiv \dfrac{3\tan\theta - \tan^3\theta}{1 - 3\tan^2\theta}$.

   b Given that θ is acute such that $\cos\theta = \frac{1}{3}$, show that
   $$\tan 3\theta = \frac{10\sqrt{2}}{23}.$$

> **Hint:** Divide formulae for $\sin 3\theta$ and $\cos 3\theta$. See Example 13 for a useful form of $\sin 3\theta$, and use a similar form for $\cos 3\theta$.

7.4 **You can write expressions of the form $a\cos\theta + b\sin\theta$, where a and b are constants, as a sine function only or a cosine function only.**

If you sketch, or draw on your calculator, the graph of $y = 3\sin x + 4\cos x$ you will see that it has the form of $y = \sin x$ or $y = \cos x$ but stretched vertically and translated horizontally.

If you draw the graph of $y = 5\sin(x + \tan^{-1}\{\frac{4}{3}\})$ or $y = 5\cos(x - \tan^{-1}\{\frac{3}{4}\})$, you will see that they are the same as $y = 3\sin x + 4\cos x$.

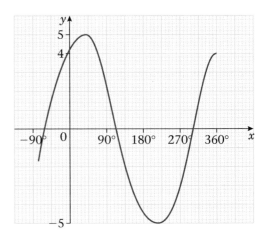

Using the addition formulae you can show that all expressions of the form $a\cos\theta + b\sin\theta$ can be expressed in one of the forms
$R\sin(x \pm \alpha)$ where $R > 0$, and $0 < \alpha < 90°$, or
$R\cos(x \pm \beta)$ where $R > 0$, and $0 < \beta < 90°$.

> **Remember:** the graph of $y = a\,f(x - \alpha)$ is the graph of $y = f(x)$ stretched vertically by a factor of a and translated horizontally by α.

Example 16

Show that you can express $3\sin x + 4\cos x$ in the form $R\sin(x + \alpha)$, where $R > 0$, $0 < \alpha < 90°$, giving your values of R and α to 1 decimal place where appropriate.

$R \sin(x + \alpha) \equiv R \sin x \cos \alpha + R \cos x \sin \alpha$

Use $\sin(A + B) \equiv \sin A \cos B + \cos A \sin B$, and multiply through by R.

Let $3 \sin x + 4 \cos x \equiv R \sin x \cos \alpha$
$\qquad\qquad\qquad\qquad + R \cos x \sin \alpha$

For this to be true for all values of x, the coefficients of $\sin x$ and $\cos x$ on both sides of the identity have to be equal.

So $R \cos \alpha = 3$ and $R \sin \alpha = 4$

Divide the equations to find $\tan \alpha$

$$\frac{R \sin \alpha}{R \cos \alpha} = \tfrac{4}{3} \text{ or } \tan \alpha = \tfrac{4}{3}$$

So $\alpha = 53.1°$ (1 d.p.)

Equations of this sort can always be solved, and so R and α can always be found.

You could draw a right-angled triangle with $\cos \alpha = \dfrac{3}{R}$ and $\sin \alpha = \dfrac{4}{R}$

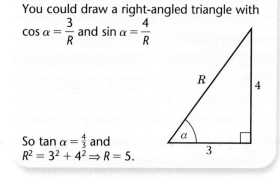

Square and add the equations to find R^2:

$$R^2 \cos^2 \alpha + R^2 \sin^2 \alpha = 3^2 + 4^2$$
So $\quad R^2(\cos^2 \alpha + \sin^2 \alpha) = 3^2 + 4^2$
So $\qquad\qquad\qquad R = \sqrt{3^2 + 4^2} = 5$

So $\tan \alpha = \tfrac{4}{3}$ and
$R^2 = 3^2 + 4^2 \Rightarrow R = 5$.

$3 \sin x + 4 \cos x \equiv 5 \sin(x + 53.1°)$

You could equally have shown that $3 \sin x + 4 \cos x \equiv 5 \cos(x - 36.9°)$ by setting $3 \sin x + 4 \cos x \equiv R \cos(x - \alpha)$ and solving for R and α, as in the example.

Example 17

a Show that you can express $\sin x - \sqrt{3} \cos x$ in the form $R \sin(x - \alpha)$, where $R > 0$, $0 < \alpha < \dfrac{\pi}{2}$.

b Hence sketch the graph of $y = \sin x - \sqrt{3} \cos x$.

a Set $\sin x - \sqrt{3} \cos x \equiv R \sin(x - \alpha)$
$\qquad \sin x - \sqrt{3} \cos x \equiv R \sin x \cos \alpha$
$\qquad\qquad\qquad\qquad\qquad - R \cos x \sin \alpha$

Expand $\sin(x - \alpha)$ and multiply by R.

So $R \cos \alpha = 1$ and $R \sin \alpha = \sqrt{3}$

Compare the coefficients of $\sin x$ and $\cos x$ on both sides of the identity.

Dividing, $\tan \alpha = \sqrt{3}$, so $\alpha = \dfrac{\pi}{3}$

Squaring and adding: $R = 2$
So $\sin x - \sqrt{3} \cos x \equiv 2 \sin\left(x - \dfrac{\pi}{3}\right)$

b $y = \sin x - \sqrt{3} \cos x \equiv 2 \sin\left(x - \dfrac{\pi}{3}\right)$

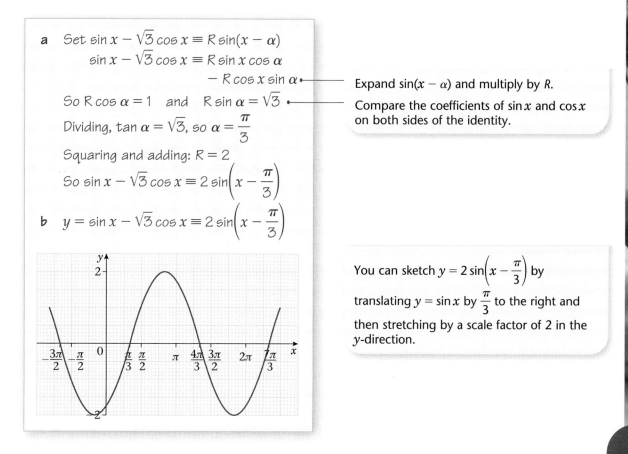

You can sketch $y = 2 \sin\left(x - \dfrac{\pi}{3}\right)$ by

translating $y = \sin x$ by $\dfrac{\pi}{3}$ to the right and

then stretching by a scale factor of 2 in the y-direction.

Example 18

a Express $2\cos\theta + 5\sin\theta$ in the form $R\cos(\theta - \alpha)$, where $R > 0$, $0 < \alpha < 90°$.

b Hence solve, for $0 < \theta < 360°$, the equation $2\cos\theta + 5\sin\theta = 3$.

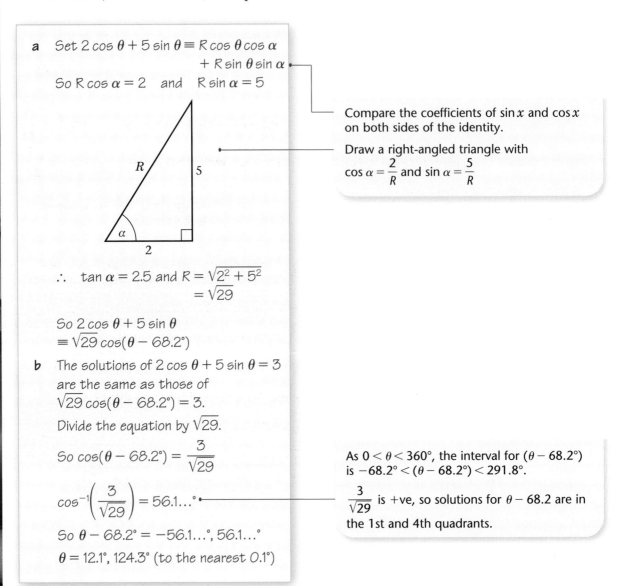

a Set $2\cos\theta + 5\sin\theta \equiv R\cos\theta\cos\alpha$
$+ R\sin\theta\sin\alpha$

So $R\cos\alpha = 2$ and $R\sin\alpha = 5$

> Compare the coefficients of $\sin x$ and $\cos x$ on both sides of the identity.

> Draw a right-angled triangle with $\cos\alpha = \dfrac{2}{R}$ and $\sin\alpha = \dfrac{5}{R}$

$\therefore \quad \tan\alpha = 2.5$ and $R = \sqrt{2^2 + 5^2}$
$= \sqrt{29}$

So $2\cos\theta + 5\sin\theta$
$\equiv \sqrt{29}\cos(\theta - 68.2°)$

b The solutions of $2\cos\theta + 5\sin\theta = 3$
are the same as those of
$\sqrt{29}\cos(\theta - 68.2°) = 3$.
Divide the equation by $\sqrt{29}$.

So $\cos(\theta - 68.2°) = \dfrac{3}{\sqrt{29}}$

$\cos^{-1}\left(\dfrac{3}{\sqrt{29}}\right) = 56.1...°$

So $\theta - 68.2° = -56.1...°, 56.1...°$

$\theta = 12.1°, 124.3°$ (to the nearest 0.1°)

> As $0 < \theta < 360°$, the interval for $(\theta - 68.2°)$ is $-68.2° < (\theta - 68.2°) < 291.8°$.

> $\dfrac{3}{\sqrt{29}}$ is +ve, so solutions for $\theta - 68.2$ are in the 1st and 4th quadrants.

Example 19

Without using calculus, find the maximum value of $12\cos\theta + 5\sin\theta$, and give the smallest positive value of θ at which it arises.

Set $12\cos\theta + 5\sin\theta \equiv R\cos(\theta - \alpha)$
So $12\cos\theta + 5\sin\theta \equiv R\cos\theta\cos\alpha$
$+ R\sin\theta\sin\alpha$
So $R\cos\alpha = 12$ and $R\sin\alpha = 5$

$R = 13$ and $\tan\alpha = \dfrac{5}{12} \Rightarrow \alpha = 22.6°$

> The most convenient forms to use here are $R\sin(\theta + \alpha)$ or $R\cos(\theta - \alpha)$ as the sign in the expanded form is the same as that in $12\cos\theta + 5\sin\theta$. If the signs do not match up, it will be more difficult for you.

So $12 \cos \theta + 5 \sin \theta \equiv 13 \cos(\theta - 22.6°)$

The maximum value of $13 \cos(\theta - 22.6°)$ is 13 and occurs when $\cos(\theta - 22.6°) = 1$; i.e. when $\theta - 22.6° = \ldots, -360°, 0°, 360°, \ldots$ The smallest positive value of θ, therefore, is 22.6°.

■ For positive values of a and b,

$a \sin \theta \pm b \cos \theta$ can be expressed in the form $R \sin(\theta \pm \alpha)$, with $R > 0$ and $0 < \alpha < 90°$ $\left(\text{or } \dfrac{\pi}{2}\right)$

$a \cos \theta \pm b \sin \theta$ can be expressed in the form $R \cos(\theta \mp \alpha)$, with $R > 0$ and $0 < \alpha < 90°$ $\left(\text{or } \dfrac{\pi}{2}\right)$

where $R \cos \alpha = a$ and $R \sin \alpha = b$

and $R = \sqrt{a^2 + b^2}$.

> Do not quote these results, but they are useful check points.

Note: When solving equations of the form $a \cos \theta + b \sin \theta = c$, use the 'R formula', unless $c = 0$, when the equation reduces to $\tan \theta = k$.

Exercise 7D

Give all angles to the nearest 0.1° and non-exact values of R in surd form.

1 Given that $5 \sin \theta + 12 \cos \theta \equiv R \sin(\theta + \alpha)$, find the value of R, $R > 0$, and the value of $\tan \alpha$.

2 Given that $\sqrt{3} \sin \theta + \sqrt{6} \cos \theta \equiv 3 \cos(\theta - \alpha)$, where $0 < \alpha < 90°$, find the value of α.

3 Given that $2 \sin \theta - \sqrt{5} \cos \theta \equiv -3 \cos(\theta + \alpha)$, where $0 < \alpha < 90°$, find the value of α.

4 Show that:

a $\cos \theta + \sin \theta \equiv \sqrt{2} \sin\left(\theta + \dfrac{\pi}{4}\right)$ **b** $\sqrt{3} \sin 2\theta - \cos 2\theta \equiv 2 \sin\left(2\theta - \dfrac{\pi}{6}\right)$

5 Prove that $\cos 2\theta - \sqrt{3} \sin 2\theta \equiv 2 \cos\left(2\theta + \dfrac{\pi}{3}\right) \equiv -2 \sin\left(2\theta - \dfrac{\pi}{6}\right)$.

6 Find the value of R, where $R > 0$, and the value of α, where $0 < \alpha < 90°$, in each of the following cases:

a $\sin \theta + 3 \cos \theta \equiv R \sin(\theta + \alpha)$ **b** $3 \sin \theta - 4 \cos \theta \equiv R \sin(\theta - \alpha)$

c $2 \cos \theta + 7 \sin \theta \equiv R \cos(\theta - \alpha)$ **d** $\cos 2\theta - 2 \sin 2\theta \equiv R \cos(2\theta + \alpha)$

7 **a** Show that $\cos \theta - \sqrt{3} \sin \theta$ can be written in the form $R \cos(\theta + \alpha)$, with $R > 0$ and $0 < \alpha < \dfrac{\pi}{2}$.

b Hence sketch the graph of $y = \cos \theta - \sqrt{3} \sin \theta$, $0 < \alpha < 2\pi$, giving the coordinates of points of intersection with the axes.

8 **a** Show that $3 \sin 3\theta - 4 \cos 3\theta$ can be written in the form $R \sin(3\theta - \alpha)$, with $R > 0$ and $0 < \alpha < 90°$.

b Deduce the minimum value of $3 \sin 3\theta - 4 \cos 3\theta$ and work out the smallest positive value of θ at which it occurs.

9 **a** Show that $\cos 2\theta + \sin 2\theta$ can be written in the form $R\sin(2\theta + \alpha)$, with $R > 0$ and $0 < \alpha < \dfrac{\pi}{2}$.

b Hence solve, in the interval $0 \leqslant \theta < 2\pi$, the equation $\cos 2\theta + \sin 2\theta = 1$, giving your answers as rational multiples of π.

10 **a** Express $7\cos\theta - 24\sin\theta$ in the form $R\cos(\theta + \alpha)$, with $R > 0$ and $0 < \alpha < 90°$.

b The graph of $y = 7\cos\theta - 24\sin\theta$ meets the y-axis at P. State the coordinates of P.

c Write down the maximum and minimum values of $7\cos\theta - 24\sin\theta$.

d Deduce the number of solutions, in the interval $0 < \theta < 360°$, of the following equations:
i $7\cos\theta - 24\sin\theta = 15$ **ii** $7\cos\theta - 24\sin\theta = 26$ **iii** $7\cos\theta - 24\sin\theta = -25$

11 **a** Express $5\sin^2\theta - 3\cos^2\theta + 6\sin\theta\cos\theta$ in the form $a\sin 2\theta + b\cos 2\theta + c$, where a, b and c are constants.

b Hence find the maximum and minimum values of $5\sin^2\theta - 3\cos^2\theta + 6\sin\theta\cos\theta$.

12 Solve the following equations, in the interval given in brackets:
a $6\sin x + 8\cos x = 5\sqrt{3}$ $[0, 360°]$
b $2\cos 3\theta - 3\sin 3\theta = -1$ $[0, 90°]$
c $8\cos\theta + 15\sin\theta = 10$ $[0, 360°]$
d $5\sin\dfrac{x}{2} - 12\cos\dfrac{x}{2} = -6.5$ $[-360°, 360°]$

13 Solve the following equations, in the interval given in brackets:
a $\sin x\cos x = 1 - 2.5\cos 2x$ $[0, 360°]$
b $\cot\theta + 2 = \csc\theta$ $[0 < \theta < 360°, \theta \neq 180°]$
c $\sin\theta = 2\cos\theta - \sec\theta$ $[0, 180°]$
d $\sqrt{2}\cos\left(\theta - \dfrac{\pi}{4}\right) + (\sqrt{3} - 1)\sin\theta = 2$ $[0, 2\pi]$

14 Solve, if possible, in the interval $0 < \theta < 360°$, $\theta \neq 180°$, the equation $\dfrac{4 - 2\sqrt{2}\sin\theta}{1 + \cos\theta} = k$ in the case when k is equal to:
a 4 **b** 2 **c** 1 **d** 0 **e** -1

15 A class were asked to solve $3\cos\theta = 2 - \sin\theta$ for $0 \leqslant \theta \leqslant 360°$. One student expressed the equation in the form $R\cos(\theta - \alpha) = 2$, with $R > 0$ and $0 < \alpha < 90°$, and correctly solved the equation.

a Find the values of R and α and hence find her solutions.

Another student decided to square both sides of the equation and then form a quadratic equation in $\sin\theta$.

b Show that the correct quadratic equation is $10\sin^2\theta - 4\sin\theta - 5 = 0$.

c Solve this equation, for $0 \leqslant \theta < 360°$.

d Explain why not all of the answers satisfy $3\cos\theta = 2 - \sin\theta$.

7.5 You can express sums and differences of sines and cosines as products of sines and/or cosines by using the 'factor formulae'.

- $\sin P + \sin Q \equiv 2\sin\left(\dfrac{P+Q}{2}\right)\cos\left(\dfrac{P-Q}{2}\right)$
- $\cos P + \cos Q \equiv 2\cos\left(\dfrac{P+Q}{2}\right)\cos\left(\dfrac{P-Q}{2}\right)$
- $\sin P - \sin Q \equiv 2\cos\left(\dfrac{P+Q}{2}\right)\sin\left(\dfrac{P-Q}{2}\right)$
- $\cos P - \cos Q \equiv -2\sin\left(\dfrac{P+Q}{2}\right)\sin\left(\dfrac{P-Q}{2}\right)$

These identities are derived from the addition formulae.

Example 20

Use the formulae for $\sin(A + B)$ and $\sin(A - B)$ to derive the result that

$$\sin P + \sin Q \equiv 2 \sin\left(\frac{P + Q}{2}\right) \cos\left(\frac{P - Q}{2}\right)$$

$\sin(A + B) \equiv \sin A \cos B + \cos A \sin B$
and $\sin(A - B) \equiv \sin A \cos B - \cos A \sin B$

Add the two identities:

$\sin(A + B) + \sin(A - B) \equiv 2 \sin A \cos B$

Let $A + B = P$ and $A - B = Q$,

then $A = \dfrac{P + Q}{2}$ and $B = \dfrac{P - Q}{2}$

$\sin P + \sin Q \equiv 2 \sin\left(\dfrac{P + Q}{2}\right) \cos\left(\dfrac{P - Q}{2}\right)$

> The other three factor formulae are proved in a similar manner, by adding or subtracting two appropriate addition formulae. See Exercise 7E.

> This result is useful in integration, e.g.
> $\displaystyle\int 2 \sin 4x \cos x \, dx = \int (\sin 5x + \sin 3x) dx.$

Example 21

Using the result that $\sin P - \sin Q \equiv 2 \cos\left(\dfrac{P + Q}{2}\right) \sin\left(\dfrac{P - Q}{2}\right)$

a show that $\sin 105° - \sin 15° = \dfrac{1}{\sqrt{2}}$

b solve, for $0 \leqslant \theta \leqslant \pi$, $\sin 4\theta - \sin 3\theta = 0$

a $\sin 105° - \sin 15°$

$= 2 \cos\left(\dfrac{105° + 15°}{2}\right) \sin\left(\dfrac{105° - 15°}{2}\right)$

$= 2 \cos 60° \sin 45°$

$= 2\left(\dfrac{1}{2}\right)\left(\dfrac{1}{\sqrt{2}}\right)$

$= \dfrac{1}{\sqrt{2}}$

> Let $P = 105°$ and $Q = 15°$.

> Remember: $\cos 60° = \dfrac{1}{2}$, $\sin 45° = \dfrac{1}{\sqrt{2}}$.

b $\quad \sin 4\theta - \sin 3\theta = 2\cos\left(\dfrac{7\theta}{2}\right)\sin\left(\dfrac{\theta}{2}\right)$ —— Let $P = 4\theta$ and $Q = 3\theta$.

The solutions of $2\cos\left(\dfrac{7\theta}{2}\right)\sin\left(\dfrac{\theta}{2}\right) = 0$
are either

$$\cos\left(\dfrac{7\theta}{2}\right) = 0$$ —— As $0 \leqslant \theta \leqslant \pi$, the interval for $\dfrac{7\theta}{2}$
is $0 \leqslant \dfrac{7\theta}{2} \leqslant \dfrac{7\pi}{2}$.

so $\quad \dfrac{7\theta}{2} = \dfrac{\pi}{2}, \dfrac{3\pi}{2}, \dfrac{5\pi}{2}, \dfrac{7\pi}{2}$

$\therefore \quad \theta = \dfrac{\pi}{7}, \dfrac{3\pi}{7}, \dfrac{5\pi}{7}, \pi$

or $\quad \sin\left(\dfrac{\theta}{2}\right) = 0$ —— The interval for $\dfrac{\theta}{2}$ is $0 \leqslant \dfrac{\theta}{2} \leqslant \dfrac{\pi}{2}$.

so $\quad \dfrac{\theta}{2} = 0 \quad \therefore \quad \theta = 0$

Answers are $\theta = 0, \dfrac{\pi}{7}, \dfrac{3\pi}{7}, \dfrac{5\pi}{7}, \pi$

Example 22

Prove that $\dfrac{\sin(x + 2y) + \sin(x + y) + \sin x}{\cos(x + 2y) + \cos(x + y) + \cos x} \equiv \tan(x + y)$.

In the numerator

$\sin(x + 2y) + \sin x$

$\equiv 2\sin\left(\dfrac{2x + 2y}{2}\right)\cos\left(\dfrac{2y}{2}\right)$ —— Use $\sin P + \sin Q \equiv 2\sin\left(\dfrac{P + Q}{2}\right)\cos\left(\dfrac{P - Q}{2}\right)$ with $P = x + 2y$ and $Q = x$.

$\equiv 2\sin(x + y)\cos y$

So $\sin(x + 2y) + \sin(x + y) + \sin x$
$\equiv \sin(x + y) + 2\sin(x + y)\cos y$
$\equiv \sin(x + y)(1 + 2\cos y)$ ① —— Factorise.

Similarly for the denominator

$\cos(x + 2y) + \cos(x + y) + \cos x$ —— Use $\cos P + \cos Q \equiv 2\cos\left(\dfrac{P + Q}{2}\right)\cos\left(\dfrac{P - Q}{2}\right)$ with $P = x + 2y$ and $Q = x$.
$\equiv \cos(x + y) + 2\cos(x + y)\cos y$
$\equiv \cos(x + y)(1 + 2\cos y)$ ② —— Factorise.

so $\dfrac{\sin(x + 2y) + \sin(x + y) + \sin x}{\cos(x + 2y) + \cos(x + y) + \cos x}$ —— Use results ① and ②.

$\equiv \dfrac{\sin(x + y)\cancel{(1 + 2\cos y)}}{\cos(x + y)\cancel{(1 + 2\cos y)}}$ —— Cancel.

$\equiv \tan(x + y)$

Exercise 7E

1 a Show that $\sin(A + B) + \sin(A - B) \equiv 2\sin A \cos B$.

b Deduce that $\sin P + \sin Q \equiv 2 \sin\left(\dfrac{P + Q}{2}\right)\cos\left(\dfrac{P - Q}{2}\right)$.

[handwritten: let $P = A+B$ $Q = A - B$]

[handwritten: $\sin P + \sin Q = 2\sin\left(\dfrac{P+Q}{2}\right)\cos\left(\dfrac{P-Q}{2}\right)$]

c Use part **a** to express the following as the sum of two sines:

   i $2 \sin 7\theta \cos 2\theta$ **ii** $2 \sin 12\theta \cos 5\theta$

[handwritten: $P+Q = 24\theta$ $P - Q = 10\theta$ $2Q = 14\theta$]

d Use the result in **b** to solve, in the interval $0 \le \theta \le 180°$, $\sin 3\theta + \sin \theta = 0$.

e Prove that $\dfrac{\sin 7\theta + \sin \theta}{\sin 5\theta + \sin 3\theta} \equiv \dfrac{\cos 3\theta}{\cos \theta}$.

[handwritten: $\sin 2\theta = 0$ $\sin \theta = 0$]
[handwritten: $\theta = 0, 180, 90$]

2 a Show that $\sin(A + B) - \sin(A - B) \equiv 2\cos A \sin B$.

b Express the following as the difference of two sines:

   i $2\cos 5x \sin 3x$ **ii** $\cos 2x \sin x$ **iii** $6\cos \frac{3}{2}x \sin \frac{1}{2}x$

c Using the result in **a** show that $\sin P - \sin Q \equiv 2 \cos\left(\dfrac{P + Q}{2}\right)\sin\left(\dfrac{P - Q}{2}\right)$.

d Deduce that $\sin 56° - \sin 34° = \sqrt{2} \sin 11°$.

3 a Show that $\cos(A + B) + \cos(A - B) \equiv 2\cos A \cos B$.

b Express as a sum of cosines   **i** $2\cos \dfrac{5\theta}{2} \cos \dfrac{\theta}{2}$ **ii** $5 \cos 2x \cos 3x$

c Show that $\cos P + \cos Q \equiv 2 \cos\left(\dfrac{P + Q}{2}\right)\cos\left(\dfrac{P - Q}{2}\right)$.

d Prove that $\dfrac{\sin 3\theta - \sin \theta}{\cos 3\theta + \cos \theta} \equiv \tan \theta$.

4 a Show that $\cos(A + B) - \cos(A - B) \equiv -2\sin A \sin B$.

b Hence show that $\cos P - \cos Q \equiv -2\sin\left(\dfrac{P + Q}{2}\right)\sin\left(\dfrac{P - Q}{2}\right)$.

c Deduce that $\cos 2\theta - 1 \equiv -2 \sin^2 \theta$.

> **Hint** to Question 4c:
> What is the value of $\cos 0°$?

d Solve, in the interval $0 \le \theta \le 180°$, $\cos 3\theta + \sin 2\theta - \cos \theta = 0$.

5 Express the following as a sum or difference of sines or cosines:

   a $2 \sin 8x \cos 2x$ **b** $\cos 5x \cos x$ **c** $3 \sin x \sin 7x$ *[handwritten: $= \frac{3}{2}(\cos 8x - \cos 6x)$]*

[handwritten: $16x = P+Q$ $4x = P - Q$]
[handwritten: $\sin 5x + \sin 10x$]

[handwritten: $P - Q = 80$]
[handwritten: $P + Q = 200$]
   d $\cos 100° \cos 40°$ *[handwritten: 140 $\frac{1}{2}(\cos 80 + \cos 140)$]*   **e** $10 \cos \dfrac{3x}{2} \sin \dfrac{x}{2}$ **f** $2 \sin 30° \cos 10°$

6 Show, without using a calculator, that $2 \sin 82\frac{1}{2}° \cos 37\frac{1}{2}° = \frac{1}{2}(\sqrt{3} + \sqrt{2})$.

7 Express, in their simplest form, as a product of sines and/or cosines:

   a $\sin 12x + \sin 8x$ **b** $\cos(x + 2y) - \cos(2y - x)$ **c** $(\cos 4x + \cos 2x)\sin x$

   d $\sin 95° - \sin 5°$ **e** $\cos \dfrac{\pi}{15} + \cos \dfrac{\pi}{12}$ **f** $\sin 150° + \sin 20°$

8 Using the identity $\cos P + \cos Q \equiv 2 \cos\left(\dfrac{P + Q}{2}\right)\cos\left(\dfrac{P - Q}{2}\right)$, show that

   $\cos \theta + \cos\left(\theta + \dfrac{2\pi}{3}\right) + \cos\left(\theta + \dfrac{4\pi}{3}\right) = 0$.

9 Prove that $\dfrac{\sin 75° + \sin 15°}{\cos 15° - \cos 75°} = \sqrt{3}$.

10 Solve the following equations:

a $\cos 4x = \cos 2x$, for $0 \leqslant x \leqslant 180°$

b $\sin 3\theta - \sin \theta = 0$, for $0 \leqslant \theta \leqslant 2\pi$

c $\sin(x + 20°) + \sin(x - 10°) = \cos 15°$, for $0 \leqslant x \leqslant 360°$

d $\sin 3\theta - \sin \theta = \cos 2\theta$, for $0 \leqslant \theta \leqslant 2\pi$

11 Prove the identities

a $\dfrac{\sin 7\theta - \sin 3\theta}{\sin \theta \cos \theta} \equiv 4 \cos 5\theta$

b $\dfrac{\cos 2\theta + \cos 4\theta}{\sin 2\theta - \sin 4\theta} \equiv -\cot \theta$

c $\sin^2(x + y) - \sin^2(x - y) \equiv \sin 2x \sin 2y$

d $\cos x + 2 \cos 3x + \cos 5x \equiv 4 \cos^2 x \cos 3x$

12 a Prove that $\cos \theta + \sin 2\theta - \cos 3\theta \equiv \sin 2\theta(1 + 2 \sin \theta)$.

b Hence solve, for $0 \leqslant \theta \leqslant 2\pi$, $\cos \theta + \sin 2\theta = \cos 3\theta$.

Mixed exercise 7F

1 The lines l_1 and l_2, with equations $y = 2x$ and $3y = x - 1$ respectively, are drawn on the same set of axes. Given that the scales are the same on both axes and that the angles that l_1 and l_2 make with the positive x-axis are A and B respectively,

a write down the value of $\tan A$ and the value of $\tan B$;

b without using your calculator, work out the acute angle between l_1 and l_2.

2 Given that $\sin x = \dfrac{1}{\sqrt{5}}$ where x is acute, and that $\cos(x - y) = \sin y$, show that $\tan y = \dfrac{\sqrt{5} + 1}{2}$.

3 Using $\tan 2\theta = \dfrac{2 \tan \theta}{1 - \tan^2 \theta}$ with an appropriate value of θ,

a show that $\tan \dfrac{\pi}{8} = \sqrt{2} - 1$.

b Use the result in **a** to find the exact value of $\tan \dfrac{3\pi}{8}$.

4 In $\triangle ABC$, $AB = 5$ cm and $AC = 4$ cm, $\angle ABC = (\theta - 30)°$ and $\angle ACB = (\theta + 30)°$. Using the sine rule, show that $\tan \theta = 3\sqrt{3}$.

5 Two of the angles, A and B, in $\triangle ABC$ are such that $\tan A = \frac{3}{4}$, $\tan B = \frac{5}{12}$.

a Find the exact value of **i** $\sin(A + B)$ **ii** $\tan 2B$

b By writing C as $180° - (A + B)$, show that $\cos C = -\frac{33}{65}$.

6 Show that

a $\sec \theta \operatorname{cosec} \theta \equiv 2 \operatorname{cosec} 2\theta$

b $\dfrac{1 - \cos 2x}{1 + \cos 2x} \equiv \sec^2 x - 1$

c $\cot \theta - 2 \cot 2\theta \equiv \tan \theta$

d $\cos^4 2\theta - \sin^4 2\theta \equiv \cos 4\theta$

e $\tan\left(\dfrac{\pi}{4} + x\right) - \tan\left(\dfrac{\pi}{4} - x\right) \equiv 2 \tan 2x$

f $\sin(x + y) \sin(x - y) \equiv \cos^2 y - \cos^2 x$

g $1 + 2 \cos 2\theta + \cos 4\theta \equiv 4 \cos^2 \theta \cos 2\theta$

7 The angles x and y are acute angles such that $\sin x = \dfrac{2}{\sqrt{5}}$ and $\cos y = \dfrac{3}{\sqrt{10}}$.

  a Show that $\cos 2x = -\frac{3}{5}$.

  b Find the value of $\cos 2y$.

  c Show without using your calculator, that

    i $\tan(x + y) = 7$ **ii** $x - y = \dfrac{\pi}{4}$

8 Given that $\sin x \cos y = \frac{1}{2}$ and $\cos x \sin y = \frac{1}{3}$,

  a show that $\sin(x + y) = 5\sin(x - y)$.

  Given also that $\tan y = k$, express in terms of k:

  b $\tan x$

  c $\tan 2x$

9 Solve the following equations in the interval given in brackets:

  a $\sqrt{3}\sin 2\theta + 2\sin^2\theta = 1$ $\{0 \leqslant \theta \leqslant \pi\}$

  b $\sin 3\theta \cos 2\theta = \sin 2\theta \cos 3\theta$ $\{0 \leqslant \theta \leqslant 2\pi\}$

  c $\sin(\theta + 40°) + \sin(\theta + 50°) = 0$ $\{0 \leqslant \theta \leqslant 360°\}$

  d $\sin^2\dfrac{\theta}{2} = 2\sin\theta$ $\{0 \leqslant \theta \leqslant 360°\}$

  e $2\sin\theta = 1 + 3\cos\theta$ $\{0 \leqslant \theta \leqslant 360°\}$

  f $\cos 5\theta = \cos 3\theta$ $\{0 \leqslant \theta \leqslant \pi\}$

  g $\cos 2\theta = 5\sin\theta$ $\{-\pi \leqslant \theta \leqslant \pi\}$.

10 The first three terms of an arithmetic series are $\sqrt{3}\cos\theta$, $\sin(\theta - 30°)$ and $\sin\theta$, where θ is acute. Find the value of θ.

11 Solve, for $0 \leqslant \theta \leqslant 360°$, $\cos(\theta + 40°)\cos(\theta - 10°) = 0.5$.

12 Without using calculus, find the maximum and minimum value of the following expressions. In each case give the smallest positive value of θ at which each occurs.

  a $\sin\theta\cos 10° - \cos\theta\sin 10°$

  b $\cos 30°\cos\theta - \sin 30°\sin\theta$

  c $\sin\theta + \cos\theta$

13 **a** Express $\sin x - \sqrt{3}\cos x$ in the form $R\sin(x - \alpha)$, with $R > 0$ and $0 < \alpha < 90°$.

  b Hence sketch the graph of $y = \sin x - \sqrt{3}\cos x$ $\{-360° \leqslant x \leqslant 360°\}$, giving the coordinates of all points of intersection with the axes.

14 Given that $7\cos 2\theta + 24\sin 2\theta \equiv R\cos(2\theta - \alpha)$, where $R > 0$ and $0 < \alpha < \dfrac{\pi}{2}$, find:

  a the value of R and the value of α, to 2 decimal places

  b the maximum value of $14\cos^2\theta + 48\sin\theta\cos\theta$

15 **a** Given that α is acute and $\tan\alpha = \frac{3}{4}$, prove that

$$3\sin(\theta + \alpha) + 4\cos(\theta + \alpha) \equiv 5\cos\theta$$

  b Given that $\sin x = 0.6$ and $\cos x = -0.8$, evaluate $\cos(x + 270)°$ and $\cos(x + 540)°$.   Ⓔ

16 a Without using a calculator, find the values of:

i $\sin 40° \cos 10° - \cos 40° \sin 10°$ **ii** $\dfrac{1}{\sqrt{2}} \cos 15° - \dfrac{1}{\sqrt{2}} \sin 15°$ **iii** $\dfrac{1 - \tan 15°}{1 + \tan 15°}$

b Find, to 1 decimal place, the values of x, $0 \leqslant x \leqslant 360°$, which satisfy the equation

$2 \sin x = \cos(x - 60)$ **E**

17 a Prove, by counter example, that the statement

'$\sec(A + B) \equiv \sec A + \sec B$, for all A and B'

is false.

b Prove that $\tan \theta + \cot \theta \equiv 2 \operatorname{cosec} 2\theta$, $\theta \neq \dfrac{n\pi}{2}$, $n \in \mathbb{Z}$. **E**

18 Using the formula $\cos(A + B) \equiv \cos A \cos B - \sin A \sin B$:

a Show that $\cos(A - B) - \cos(A + B) \equiv 2 \sin A \sin B$.

b Hence show that $\cos 2x - \cos 4x \equiv 2 \sin 3x \sin x$.

c Find all solutions in the range $0 \leqslant x \leqslant \pi$ of the equation

$\cos 2x - \cos 4x = \sin x$

giving all your solutions in multiples of π radians. **E**

19 a Given that $\cos(x + 30°) = 3 \cos(x - 30°)$, prove that $\tan x = -\dfrac{\sqrt{3}}{2}$.

b i Prove that $\dfrac{1 - \cos 2\theta}{\sin 2\theta} = \tan \theta$.

ii Verify that $\theta = 180°$ is a solution of the equation $\sin 2\theta = 2 - 2 \cos 2\theta$.

iii Using the result in part **i**, or otherwise, find the two other solutions, $0 < \theta < 360°$, of the equation $\sin 2\theta = 2 - 2 \cos 2\theta$. **E**

20 a Express $1.5 \sin 2x + 2 \cos 2x$ in the form $R \sin(2x + \alpha)$, where $R > 0$ and $0 < \alpha < \dfrac{\pi}{2}$, giving your values of R and α to 3 decimal places where appropriate.

b Express $3 \sin x \cos x + 4 \cos^2 x$ in the form $a \sin 2x + b \cos 2x + c$, where a, b and c are constants to be found.

c Hence, using your answer to part **a**, deduce the maximum value of $3 \sin x \cos x + 4 \cos^2 x$. **E**

Summary of key points

1 The addition (or compound angle) formulae are

- $\sin(A + B) \equiv \sin A \cos B + \cos A \sin B$ $\sin(A - B) \equiv \sin A \cos B - \cos A \sin B$
- $\cos(A + B) \equiv \cos A \cos B - \sin A \sin B$ $\cos(A - B) \equiv \cos A \cos B + \sin A \sin B$
- $\tan(A + B) \equiv \dfrac{\tan A + \tan B}{1 - \tan A \tan B}$ $\tan(A - B) \equiv \dfrac{\tan A - \tan B}{1 + \tan A \tan B}$

2 The double angle formulae are

- $\sin 2A \equiv 2 \sin A \cos A$
- $\cos 2A \equiv \cos^2 A - \sin^2 A \equiv 2 \cos^2 A - 1 \equiv 1 - 2 \sin^2 A$
- $\tan 2A \equiv \dfrac{2 \tan A}{1 - \tan^2 A}$

3 Expressions of the form $a \sin \theta + b \cos \theta$ can be rewritten in terms of a sine only or a cosine only, as follows:

For positive values of a and b,

$a \sin \theta \pm b \cos \theta \equiv R \sin(\theta \pm \alpha)$, with $R > 0$ and $0 < \alpha < 90°$,

$a \cos \theta \pm b \sin \theta \equiv R \cos(\theta \mp \alpha)$, with $R > 0$ and $0 < \alpha < 90°$

where $R \cos \alpha = a$, $R \sin \alpha = b$ and $R = \sqrt{a^2 + b^2}$.

> Remember you can always use 'the R formula' to solve equations of the form $a \cos \theta + b \sin \theta = c$, where a, b and c are constants, but if $c = 0$, the equation reduces to the form $\tan \theta = k$.

4 Products of sines and/or cosines can be expressed as the sum or difference of sines or cosines, using the formulae:

$2 \sin A \cos B \equiv \sin(A + B) + \sin(A - B)$ $2 \cos A \cos B \equiv \cos(A + B) + \cos(A - B)$

$2 \cos A \sin B \equiv \sin(A + B) - \sin(A - B)$ $2 \sin A \sin B \equiv -[\cos(A + B) - \cos(A - B)]$

5 Sums or differences of sines or cosines can be expressed as a product of sines and/or cosines, using 'the factor formulae':

$\sin P + \sin Q \equiv 2 \sin\left(\dfrac{P + Q}{2}\right) \cos\left(\dfrac{P - Q}{2}\right)$ $\cos P + \cos Q \equiv 2 \cos\left(\dfrac{P + Q}{2}\right) \cos\left(\dfrac{P - Q}{2}\right)$

$\sin P - \sin Q \equiv 2 \cos\left(\dfrac{P + Q}{2}\right) \sin\left(\dfrac{P - Q}{2}\right)$ $\cos P - \cos Q \equiv -2 \sin\left(\dfrac{P + Q}{2}\right) \sin\left(\dfrac{P - Q}{2}\right)$

8

Differentiation

After completing this chapter you should be able to

1 differentiate a composite function using the chain rule

2 differentiate functions that are multiplied together by using the product rule

3 differentiate rational functions using the quotient rule

4 differentiate variations on the functions of e^x and $\ln(x)$

5 differentiate variations on the functions $\sin x$, $\cos x$ and $\tan x$.

Differentiating enables you to find the gradient of a curve. In this example we could calculate how quickly the tide is rising at any given time.

This chapter allows you to explore in greater detail some of the real life examples mentioned in this, and earlier books.

For example it was mentioned in Core 2 that the rise and fall of a tide can be modelled by a trigonometric graph.

8.1 **You need to be able to differentiate a function of a function, using the chain rule.**

■ The chain rule enables you to differentiate a function of a function. In general,

- if $y = [f(x)]^n$ then $\dfrac{dy}{dx} = n[f(x)]^{n-1} f'(x)$

- if $y = f[g(x)]$ then $\dfrac{dy}{dx} = f'[g(x)]g'(x)$

> You should learn these results.

Example 1

Given that $y = (3x^4 + x)^5$ find $\dfrac{dy}{dx}$ using the chain rule.

Here $f(x) = 3x^4 + x$
So $f'(x) = 12x^3 + 1$

Using the chain rule

$\dfrac{dy}{dx} = 5(3x^4 + x)^4(12x^3 + 1)$ •——————— This uses the chain rule with $n = 5$.

$\quad = 5(12x^3 + 1)(3x^4 + x)^4$

Example 2

Given that $y = \sqrt{5x^2 + 1}$ find the value of $\dfrac{dy}{dx}$ at $(4, 9)$.

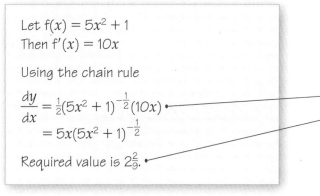

Let $f(x) = 5x^2 + 1$
Then $f'(x) = 10x$

Using the chain rule

$\dfrac{dy}{dx} = \frac{1}{2}(5x^2 + 1)^{-\frac{1}{2}}(10x)$ •——— This time $n = \frac{1}{2}$ and $\frac{1}{2}(10x)$ is simplified to $5x$.

$\quad = 5x(5x^2 + 1)^{-\frac{1}{2}}$ •——— Substitute $x = 4$ to give the required value.

Required value is $2\frac{2}{9}$. •———

■ Another form of the chain rule is

- $$\frac{dy}{dx} = \frac{dy}{du} \times \frac{du}{dx}$$

where y is a function of u, and u is a function of x.

Example 3

Given that $y = (x^2 - 7x)^4$ find $\frac{dy}{dx}$ using the chain rule.

Let $u = x^2 - 7x$, then $y = u^4$

$\therefore \dfrac{du}{dx} = 2x - 7$ and $\dfrac{dy}{du} = 4u^3$

Then, using the chain rule,

$\dfrac{dy}{dx} = \dfrac{dy}{du} \times \dfrac{du}{dx}$

$\quad = 4u^3 \times (2x - 7)$

$\quad = 4(2x - 7)(x^2 - 7x)^3$

> When the substitution is not given in a question you should put the bracket equal to u.

> Use the chain rule to find $\dfrac{dy}{dx}$.

> Ensure that you give your answer in terms of x, with no u terms present.

Example 4

Given that $y = \dfrac{1}{\sqrt{6x - 3}}$ find the value of $\dfrac{dy}{dx}$ at $(2, \frac{1}{3})$.

Let $u = 6x - 3$, then $y = u^{-\frac{1}{2}}$

$\therefore \dfrac{du}{dx} = 6$ and $\dfrac{dy}{du} = -\frac{1}{2}u^{-\frac{3}{2}}$

Then, as $\dfrac{dy}{dx} = \dfrac{dy}{du} \times \dfrac{du}{dx}$

$\dfrac{dy}{dx} = -\frac{1}{2}u^{-\frac{3}{2}} \times 6$

$\quad = -3(6x - 3)^{-\frac{3}{2}}$

Required value is $-\frac{1}{9}$.

> Put u equal to the expression in the bracket.

> Substitute $x = 2$ to give the required value.

■ Also a particular case of the chain rule is the result

- $$\frac{dy}{dx} = \frac{1}{\left(\dfrac{dx}{dy}\right)}$$

This arises since $\dfrac{dy}{dx} \times \dfrac{dx}{dy} = \dfrac{dy}{dy} = 1$

> Then you make $\dfrac{dy}{dx}$ the subject of the formula.

Example 5

Find the value of $\dfrac{dy}{dx}$ at the point (2, 1) on the curve with equation $y^3 + y = x$.

$$\frac{dx}{dy} = 3y^2 + 1$$

Start with $x = y^3 + y$ and differentiate with respect to y.

$$\therefore \frac{dy}{dx} = \frac{1}{3y^2 + 1}$$

Use $\dfrac{dy}{dx} = \dfrac{1}{\frac{dx}{dy}}$

$$= \tfrac{1}{4}$$

Substitute $y = 1$.

Exercise 8A

1 Differentiate:

a $(1 + 2x)^4$ $\quad 4(1+2x)^3 \times 2$

b $(3 - 2x^2)^{-5}$ $\quad -5(3-2x^2)^{-6} \times 11x$

c $(3 + 4x)^{\frac{1}{2}}$ $\quad \frac{1}{2}(3+4x)^{-\frac{1}{2}} \times 4$

d $(6x + x^2)^7$ $\quad 7(6x+x^2)^6$

$(6+2x)$

e $\dfrac{1}{3 + 2x}$ $\quad -1(3+2x)^{-2} \times 2$

f $\sqrt{7 - x}$ $\quad \frac{1}{2}(7-x)^{-\frac{1}{2}} \times -1$

g $4(2 + 8x)^4$ $\quad 16(2+8x)^3 \times (8)$

h $3(8 - x)^{-6}$

$-18(8-x)^{-7} \times (-1)$

2 Given that $y = \dfrac{1}{(4x + 1)^2}$ find the value of $\dfrac{dy}{dx}$ at $(\tfrac{1}{4}, \tfrac{1}{4})$.

$-2(4x+1)^{-3} \times (4) = -1$

3 Given that $y = (5 - 2x)^3$ find the value of $\dfrac{dy}{dx}$ at (1, 27).

$3(5-2x)^2 \times (-2) = -54$

4 Find the value of $\dfrac{dy}{dx}$ at the point (8, 2) on the curve with equation $3y^2 - 2y = x$.

$6y - 2 \qquad 0.1$

5 Find the value of $\dfrac{dy}{dx}$ at the point $(2\tfrac{1}{2}, 4)$ on the curve with equation $y^{\frac{1}{2}} + y^{-\frac{1}{2}} = x$.

$\frac{1}{2}y^{-\frac{1}{2}} - \frac{1}{2}y^{-\frac{3}{2}} \cdot \frac{16}{3}$

8.2 You need to differentiate functions that are multiplied together, by using the product rule.

■ To differentiate the product of two functions, differentiate the first function and leave the second function alone, then differentiate the second one and leave the first function alone, then add all this together.

● If $y = uv$ then $\dfrac{dy}{dx} = u\dfrac{dv}{dx} + v\dfrac{du}{dx}$,

where u and v are both functions of x. This is called the **product rule**.

Here is a proof of this rule:

Let $y = uv$ where u and v are two functions of x. Suppose that a small increment δx in the variable x results in a small change δu in u and a small change δv in v, which in turn results in a small change δy in the variable y.

Then $\quad y + \delta y = (u + \delta u)(v + \delta v)$ $\qquad$ ①

But $\qquad y = uv$ $\qquad\qquad\qquad\qquad$ ②

Subtract ① − ②

$$\therefore \quad \delta y = (u + \delta u)(v + \delta v) - uv$$
$$= uv + u\delta v + v\delta u + \delta u\delta v - uv$$
$$= u\delta v + v\delta u + \delta u\delta v$$
$$\therefore \quad \frac{\delta y}{\delta x} = u\frac{\delta v}{\delta x} + v\frac{\delta u}{\delta x} + \frac{\delta u}{\delta x}\delta v$$

> You will *not* need to prove this result in an examination.

As $\delta x \to 0$ then $\dfrac{\delta y}{\delta x} \to \dfrac{dy}{dx}$, $\dfrac{\delta u}{\delta x} \to \dfrac{du}{dx}$ and $\dfrac{\delta v}{\delta x} \to \dfrac{dv}{dx}$.

Also $\delta v \to 0$ and thus $\dfrac{\delta u}{\delta x}\delta v \to 0$.

$$\therefore \quad \frac{dy}{dx} = u\frac{dv}{dx} + v\frac{du}{dx}$$

> You should learn this result, and learn where it is appropriate to use it.

Example 6

Given that $f(x) = x^2\sqrt{3x - 1}$, find $f'(x)$.

> Recognise that this is a product of two functions.

Let $u = x^2$ and $v = \sqrt{3x - 1} = (3x - 1)^{\frac{1}{2}}$

Then $\dfrac{du}{dx} = 2x$ and $\dfrac{dv}{dx} = 3 \times \frac{1}{2}(3x - 1)^{-\frac{1}{2}}$

> The second function is a function of a function and requires the chain rule.

Using $\dfrac{dy}{dx} = u\dfrac{dv}{dx} + v\dfrac{du}{dx}$

$f'(x) = x^2 \times \frac{3}{2}(3x - 1)^{-\frac{1}{2}} + \sqrt{3x - 1} \times 2x$

$= \dfrac{3x^2 + 12x^2 - 4x}{2\sqrt{3x - 1}}$

$= \dfrac{15x^2 - 4x}{2\sqrt{3x - 1}}$

> Collect terms to simplify, and factorise to give the final answer.

$= \dfrac{x(15x - 4)}{2\sqrt{3x - 1}}$

Exercise 8B

1 Differentiate:

a $x(1 + 3x)^5$ **b** $2x(1 + 3x^2)^3$ **c** $x^3(2x + 6)^4$ **d** $3x^2(5x - 1)^{-1}$

2 **a** Find the value of $\dfrac{dy}{dx}$ at the point $(1, 8)$ on the curve with equation $y = x^2(3x - 1)^3$.

    b Find the value of $\dfrac{dy}{dx}$ at the point $(4, 36)$ on the curve with equation $y = 3x(2x + 1)^{\frac{1}{2}}$.

    c Find the value of $\dfrac{dy}{dx}$ at the point $(2, \frac{1}{5})$ on the curve with equation $y = (x - 1)(2x + 1)^{-1}$.

3 Find the points where the gradient is zero on the curve with equation $y = (x - 2)^2(2x + 3)$.

$\frac{dy}{dx} = (x-2)^2 \times 2 + (2x+3) \times 2(x-2)^1 x$

$= 2(x-2)^2 + 2(x-2)(2x+3) = 2(x-2)(x-2+2x+3)$

8.3 You need to be able to differentiate rational functions using the quotient rule.

■ A rational function has the form $\dfrac{u(x)}{v(x)}$, where $u(x)$ and $v(x)$ are functions.

● If $y = \dfrac{u(x)}{v(x)}$ then $\dfrac{dy}{dx} = \dfrac{v\dfrac{du}{dx} - u\dfrac{dv}{dx}}{v^2}$

This is called the **quotient rule**.

Example 7

Given that $y = \dfrac{x}{2x + 5}$ find $\dfrac{dy}{dx}$.

Let $u = x$ and $v = 2x + 5$

$\dfrac{du}{dx} = 1$ and $\dfrac{dv}{dx} = 2$

Using $\dfrac{dy}{dx} = \dfrac{v\dfrac{du}{dx} - u\dfrac{dv}{dx}}{v^2}$ —————— Recognise that y is a quotient and use the quotient rule.

$\dfrac{dy}{dx} = \dfrac{(2x + 5) \times 1 - x \times 2}{(2x + 5)^2}$

$= \dfrac{5}{(2x + 5)^2}$ —————— Simplify the numerator of the fraction.

Example 8

By expressing $y = \dfrac{u}{v}$ as $y = uv^{-1}$, prove the quotient rule.

You can use the product rule to give

$\dfrac{dy}{dx} = u\dfrac{d}{dx}(v^{-1}) + v^{-1}\dfrac{du}{dx}$

$= u\left(-v^{-2}\dfrac{dv}{dx}\right) + v^{-1}\dfrac{du}{dx}$ —————— Use the chain rule to differentiate v^{-1}.

$= -\dfrac{u}{v^2}\dfrac{dv}{dx} + \dfrac{1}{v}\dfrac{du}{dx}$

$\therefore \dfrac{dy}{dx} = \dfrac{v\dfrac{du}{dx} - u\dfrac{dv}{dx}}{v^2}$ —————— Then use a common denominator v^2.

—————— You should learn this result, and learn where it is appropriate to use it.

Exercise 8C

1 Differentiate:

   a $\dfrac{5x}{x+1}$
   b $\dfrac{2x}{3x-2}$
   c $\dfrac{x+3}{2x+1}$
   d $\dfrac{3x^2}{(2x-1)^2}$
   e $\dfrac{6x}{(5x+3)^{\frac{1}{2}}}$

2 Find the value of $\dfrac{dy}{dx}$ at the point $(1, \frac{1}{4})$ on the curve with equation $y = \dfrac{x}{3x+1}$.

3 Find the value of $\dfrac{dy}{dx}$ at the point $(12, 3)$ on the curve with equation $y = \dfrac{x+3}{(2x+1)^{\frac{1}{2}}}$.

$\dfrac{dy}{dx} = \dfrac{(2x+1)^{\frac{1}{2}} \times 1 - (x+3)(2x+1)^{-\frac{1}{2}}}{2x+1} = \dfrac{(2x+1)^{\frac{1}{2}}\left(-(x+3)(2x+1)^{-1}\right)}{(2x+1)^{-\frac{1}{2}}} = \dfrac{-x+3}{2x+1}$

8.4 You need to be able to differentiate the exponential function.

In Chapter 3 you met the exponential function e^x. This is a special function because it is the only function for which $f(x) = f'(x)$.

> You should learn this result.

■ **If $y = e^x$ then $\dfrac{dy}{dx} = e^x$**

You can prove this result from first principles by the method introduced in Book C1.

Use the definition $f'(x) = \displaystyle\lim_{\delta x \to 0}\left[\dfrac{f(x + \delta x) - f(x)}{\delta x}\right]$

If $f(x) = e^x$ then $f'(x) = \displaystyle\lim_{\delta x \to 0}\left[\dfrac{e^{x+\delta x} - e^x}{\delta x}\right]$

$\qquad\qquad\qquad = \displaystyle\lim_{\delta x \to 0}\left[\dfrac{e^x(e^{\delta x} - 1)}{\delta x}\right]$

$\qquad\qquad\qquad = e^x \displaystyle\lim_{\delta x \to 0}\left[\dfrac{(e^{\delta x} - 1)}{\delta x}\right]$

The table below shows values for $\left[\dfrac{(e^{\delta x} - 1)}{\delta x}\right]$ for $e = 2.718\,282$ for progressively smaller values of δx.

	$\delta x = 0.1$	$\delta x = 0.01$	$\delta x = 0.001$	$\delta x = 0.000\,1$	$\delta x = 0.000\,01$
$\left[\dfrac{(e^{\delta x} - 1)}{\delta x}\right]$	$1.051\,709\,25$	$1.005\,016\,772$	$1.000\,500\,23$	$1.000\,050\,06$	$1.000\,005\,063$

From this table you can see that $\left[\dfrac{(e^{\delta x} - 1)}{\delta x}\right]$ approaches a limiting value of 1 as $\delta x \to 0$. This means that if $f(x) = e^x$, then $f'(x) = 1 \times e^x$. e^x is called the exponential function, where $e = 2.718\,282$ to 6 d.p. and has the property that if $y = e^x$ then $\dfrac{dy}{dx} = e^x$ also.

This result can be used together with the chain rule and the product and quotient rules to enable you to differentiate a wide range of functions. In particular:

■ **$y = e^{f(x)}$ then $\dfrac{dy}{dx} = f'(x)e^{f(x)}$**

Example 9

Differentiate **a** $5e^x$ **b** e^{2x+3} **c** xe^{x^2} **d** $\dfrac{e^{2x+3}}{x}$

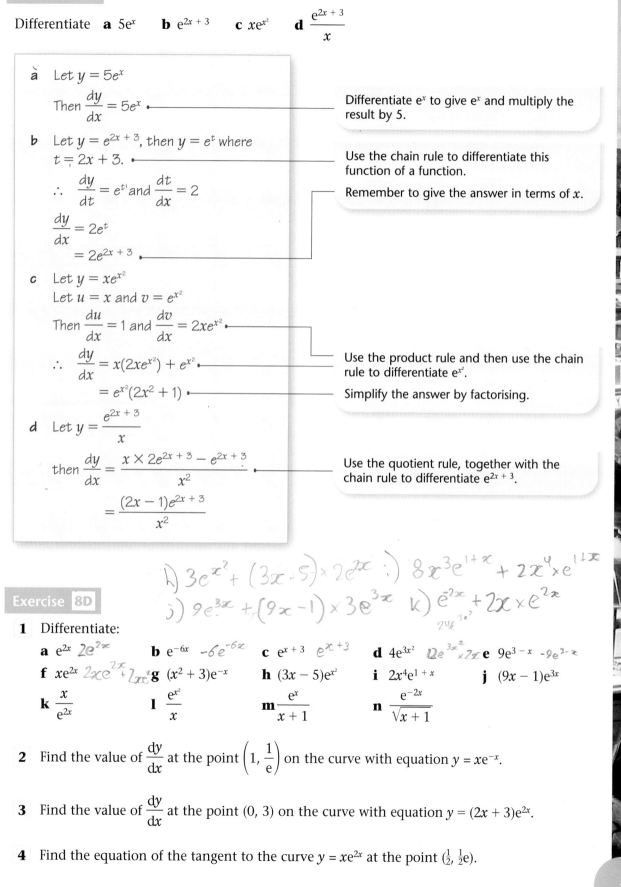

a Let $y = 5e^x$

Then $\dfrac{dy}{dx} = 5e^x$ ——————— Differentiate e^x to give e^x and multiply the result by 5.

b Let $y = e^{2x+3}$, then $y = e^t$ where $t = 2x + 3$. ——————— Use the chain rule to differentiate this function of a function.

$\therefore \dfrac{dy}{dt} = e^{t}$ and $\dfrac{dt}{dx} = 2$ ——————— Remember to give the answer in terms of x.

$\dfrac{dy}{dx} = 2e^t$

$= 2e^{2x+3}$

c Let $y = xe^{x^2}$

Let $u = x$ and $v = e^{x^2}$

Then $\dfrac{du}{dx} = 1$ and $\dfrac{dv}{dx} = 2xe^{x^2}$ ———————

$\therefore \dfrac{dy}{dx} = x(2xe^{x^2}) + e^{x^2}$ ——————— Use the product rule and then use the chain rule to differentiate e^{x^2}.

$= e^{x^2}(2x^2 + 1)$ ——————— Simplify the answer by factorising.

d Let $y = \dfrac{e^{2x+3}}{x}$

then $\dfrac{dy}{dx} = \dfrac{x \times 2e^{2x+3} - e^{2x+3}}{x^2}$ ——————— Use the quotient rule, together with the chain rule to differentiate e^{2x+3}.

$= \dfrac{(2x-1)e^{2x+3}}{x^2}$

Exercise 8D

(handwritten) h) $3e^{x^2} + (3x-5) \times 2e^{2x}$ i) $8x^3 e^{1+x} + 2x^4 x e^{11x}$

j) $9e^{3x} + (9x-1) \times 3e^{3x}$ k) $e^{2x} + 2x \times e^{2x}$

$24e^{3x^2}$

1 Differentiate:

a e^{2x} *(hw)* $2e^{2x}$ **b** e^{-6x} *(hw)* $-6e^{-6x}$ **c** e^{x+3} *(hw)* e^{x+3} **d** $4e^{3x^2}$ *(hw)* $12e^{3x^2} \times 2x$ **e** $9e^{3-x}$ *(hw)* $-9e^{3-x}$

f xe^{2x} *(hw)* $2xe^{2x} + 1xe^{2x}$ **g** $(x^2 + 3)e^{-x}$ **h** $(3x - 5)e^{x^2}$ **i** $2x^4e^{1+x}$ **j** $(9x - 1)e^{3x}$

k $\dfrac{x}{e^{2x}}$ **l** $\dfrac{e^{x^2}}{x}$ **m** $\dfrac{e^x}{x+1}$ **n** $\dfrac{e^{-2x}}{\sqrt{x+1}}$

2 Find the value of $\dfrac{dy}{dx}$ at the point $\left(1, \dfrac{1}{e}\right)$ on the curve with equation $y = xe^{-x}$.

3 Find the value of $\dfrac{dy}{dx}$ at the point $(0, 3)$ on the curve with equation $y = (2x + 3)e^{2x}$.

4 Find the equation of the tangent to the curve $y = xe^{2x}$ at the point $(\frac{1}{2}, \frac{1}{2}e)$.

5 Find the equation of the tangent to the curve $y = \dfrac{e^{\frac{x}{3}}}{x}$ at the point $(3, \frac{1}{3}e)$.

6 Find the coordinates of the turning points on the curve $y = x^2 e^{-x}$, and determine whether these points are maximum or minimum points.

7 Given that $y = \dfrac{e^{3x}}{x}$, find $\dfrac{dy}{dx}$ and $\dfrac{d^2y}{dx^2}$, simplifying your answers.

Use these answers to find the coordinates of the turning point on the curve with equation $y = \dfrac{e^{3x}}{x}$, $x > 0$, and determine the nature of this turning point.

8.5 You need to be able to differentiate the logarithmic function.

In Chapter 3 you were introduced to the logarithmic function, $\ln x$, which was defined as the inverse of the exponential function e^x.

You are now going to use the derivative of e^x to find the derivative of $\ln x$.

Let $y = \ln x$

Then $x = e^y$

So $\dfrac{dx}{dy} = e^y$

> You can make x the subject of the formula using the inverse function exp.

But $\dfrac{dy}{dx} = \dfrac{1}{\frac{dx}{dy}}$

$\therefore \dfrac{dy}{dx} = \dfrac{1}{e^y}$

$\therefore \dfrac{dy}{dx} = \dfrac{1}{x}$

> You can now use the result $\dfrac{dy}{dx} = \dfrac{1}{\frac{dx}{dy}}$, which was quoted at the end of Section 8.1, as a special case of the chain rule.

■ **So if $y = \ln x$ then $\dfrac{dy}{dx} = \dfrac{1}{x}$**

> You should learn this result.

This result can also be used together with the chain rule and the product and quotient rules to enable you to differentiate a wide range of functions.

In particular

■ **If $y = \ln[f(x)]$ then $\dfrac{dy}{dx} = \dfrac{f'(x)}{f(x)}$**

> You put $f(x) = u$, so $y = \ln u$. Then, using the chain rule, $\dfrac{dy}{dx} = \dfrac{dy}{du} \times \dfrac{du}{dx}$, so $\dfrac{dy}{dx} = \dfrac{1}{u} \times f'(x) = \dfrac{f'(x)}{f(x)}$

Example 10

Differentiate **a** $5\ln x$ **b** $\ln(6x-1)$ **c** $x^3 \ln x$ **d** $\dfrac{\ln 5x}{x}$ **e** $2x + e^x \ln x$

a $y = 5\ln x$

$\therefore \dfrac{dy}{dx} = 5 \times \dfrac{1}{x}$

$= \dfrac{5}{x}$

> If $y = af(x)$ then $\dfrac{dy}{dx} = af'(x)$.

b $y = \ln(6x - 1)$

$$\therefore \frac{dy}{dx} = 6 \times \frac{1}{6x - 1}$$

$$= \frac{6}{6x - 1}$$

> This is a function of a function. Use the chain rule.
> Let $u = 6x - 1$, so $y = \ln u$.
> $\frac{du}{dx} = 6$ and $\frac{dy}{du} = \frac{1}{u}$ $\therefore \frac{dy}{dx} = 6 \times \frac{1}{u}$.

c $y = x^3 \ln x$

$$\therefore \frac{dy}{dx} = x^3 \times \frac{1}{x} + 3x^2 \times \ln x$$

$$= x^2 + 3x^2 \ln x$$

> Use the product rule here, with $u = x^3$ and $v = \ln x$.

d $y = \frac{\ln 5x}{x}$

$$\therefore \frac{dy}{dx} = \frac{x\left(\frac{5}{5x}\right) - 1 \times \ln 5x}{x^2}$$

$$= \frac{1 - \ln 5x}{x^2}$$

> Use the quotient rule, and use the chain rule to differentiate the $\ln 5x$ term.

e $y = 2x + e^x \ln x$

$$\therefore \frac{dy}{dx} = 2 + \left(e^x \times \frac{1}{x} + e^x \ln x\right)$$

$$= 2 + \frac{e^x}{x}(1 + x \ln x)$$

> Use the product rule to differentiate the second term.

Exercise 8E

1 Find the function $f'(x)$ where $f(x)$ is

a $\ln(x + 1)$ $\frac{1}{x+1}$ **b** $\ln 2x$ $\frac{1}{x}$ **c** $\ln 3x$ $\frac{1}{x}$ **d** $\ln(5x - 4)$ $\frac{5}{5x-4}$

e $3 \ln x$ $\frac{3}{x}$ **f** $4 \ln 2x$ $\frac{4}{x}$ **g** $5 \ln(x + 4)$ $\frac{5}{x+4}$ **h** $x \ln x$ $1 + \ln x$

i $\frac{\ln x}{x + 1}$ **j** $\ln(x^2 - 5)$ **k** $(3 + x) \ln x$ **l** $e^x \ln x$

8.6 You need to be able to differentiate trigonometric functions. You can use the formula for f'(x) to differentiate sin x.

Earlier you were reminded that if you wish to differentiate a function then you must use the definition introduced in Book C1. That is,

$$f'(x) = \lim_{\delta x \to 0}\left[\frac{f(x + \delta x) - f(x)}{\delta x}\right]$$

Let $f(x) = \sin x$.

Then $f'(x) = \lim_{\delta x \to 0}\left[\frac{\sin(x + \delta x) - \sin(x)}{\delta x}\right]$

$$= \lim_{\delta x \to 0}\left[\frac{\sin x \cos \delta x + \cos x \sin \delta x - \sin x}{\delta x}\right] \quad *$$

> Now use the compound angle formula to expand $\sin(A + B)$.

As with many of these limiting values the numerator and the denominator of this fraction both approach zero, and so you need to investigate the behaviour of $\sin x$ and $\cos x$ for small values of x.

Consider first a circle with radius r, with radii AB and AC such that angle BAC is x, where x is measured in radians.

You use the formula introduced in the radians section:
area of sector $= \frac{1}{2}r^2\theta$

The area of sector ABC is $\frac{1}{2}r^2x$ and the area of triangle ABC is $\frac{1}{2}r^2\sin x$. As x becomes small the area of the triangle becomes close to the area of the sector. Thus $\frac{1}{2}r^2\sin x \approx \frac{1}{2}r^2x \Rightarrow \sin x \approx x$, where x is small and is measured in radians.

You use area of triangle $= \frac{1}{2}ab\sin C$ with $a = b = r$ and $C = x$.

Also $\cos x \approx 1$ for small values of x.

So in equation ✳ on page 141, replace $\cos \delta x$ by 1 and replace $\sin \delta x$ by δx, since δx is small.

Then $f'(x) = \lim\limits_{\delta x \to 0} \left[\dfrac{\sin x + \cos x \times \delta x - \sin x}{\delta x} \right]$

$= \cos x$

■ So if $y = \sin x$ then $\dfrac{dy}{dx} = \cos x$

This formula applies where x is measured in radians.

■ And, by the chain rule, if $y = \sin f(x)$ then $\dfrac{dy}{dx} = f'(x)\cos f(x)$

Learn these two key points.

Example 11

Differentiate **a** $y = \sin 3x$ **b** $y = \sin\frac{2}{3}x$ **c** $y = \sin^2 x$

a $y = \sin 3x$
$\dfrac{dy}{dx} = 3\cos 3x$ — Use the chain rule with $f(x) = 3x$, so $f'(x) = 3$.

b $y = \sin\frac{2}{3}x$
$\dfrac{dy}{dx} = \frac{2}{3}\cos\frac{2}{3}x$ — This time put $f(x) = \frac{2}{3}x$.

c $y = \sin^2 x = (\sin x)^2$
$\dfrac{dy}{dx} = 2(\sin x)^1 \cos x$ — Use the chain rule, with $u = \sin x$, so $\dfrac{du}{dx} = \cos x$ and $\dfrac{dy}{du} = 2u$.
$= 2\sin x \cos x$

Exercise 8F

1 Differentiate:

a $y = \sin 5x$ **b** $y = 2\sin\frac{1}{2}x$ **c** $y = 3\sin^2 x$ **d** $y = \sin(2x + 1)$

e $y = \sin 8x$ **f** $y = 6\sin\frac{2}{3}x$ **g** $y = \sin^3 x$ **h** $y = \sin^5 x$

8.7 **You can use the result obtained for the derivative of $\sin x$ to differentiate $\cos x$.**

Let $y = \cos x$

Then $y = \sin\left(\dfrac{\pi}{2} - x\right)$ •————————————— This uses the expansion of $\sin(A - B)$ together with $\sin\dfrac{\pi}{2} = 1$ and $\cos\dfrac{\pi}{2} = 0$.

Using the result that $\dfrac{dy}{dx} = f'(x)\cos f(x)$ for $y = \sin f(x)$ and

$f(x) = \dfrac{\pi}{2} - x$

$\dfrac{dy}{dx} = -\cos\left(\dfrac{\pi}{2} - x\right)$

$\quad = -\sin x$

■ So if $y = \cos x$ then $\dfrac{dy}{dx} = -\sin x$

Remember x is measured in radians.

■ Also, by the chain rule, if $y = \cos f(x)$ then $\dfrac{dy}{dx} = -f'(x)\sin f(x)$

Learn these two key points.

Example 12

Differentiate **a** $y = \cos(4x - 3)$ **b** $y = \cos x°$ (x degrees) **c** $y = \cos^3 x$

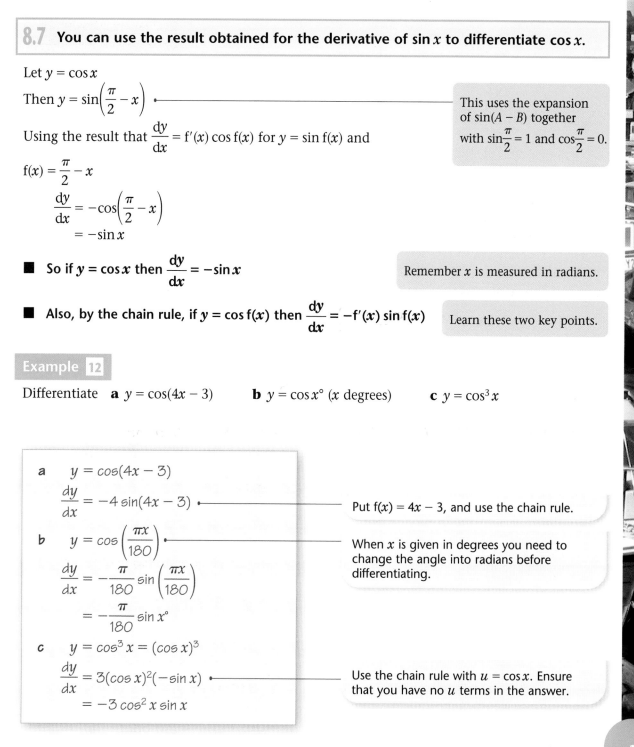

a $y = \cos(4x - 3)$

$\dfrac{dy}{dx} = -4\sin(4x - 3)$ •————————— Put $f(x) = 4x - 3$, and use the chain rule.

b $y = \cos\left(\dfrac{\pi x}{180}\right)$ •————————— When x is given in degrees you need to change the angle into radians before differentiating.

$\dfrac{dy}{dx} = -\dfrac{\pi}{180}\sin\left(\dfrac{\pi x}{180}\right)$

$\quad = -\dfrac{\pi}{180}\sin x°$

c $y = \cos^3 x = (\cos x)^3$

$\dfrac{dy}{dx} = 3(\cos x)^2(-\sin x)$ •————————— Use the chain rule with $u = \cos x$. Ensure that you have no u terms in the answer.

$\quad = -3\cos^2 x \sin x$

Exercise 8G

1 Differentiate:

a $y = 2\cos x$ b $y = \cos^5 x$ c $y = 6\cos\frac{5}{6}x$ d $y = 4\cos(3x + 2)$

e $y = \cos 4x$ f $y = 3\cos^2 x$ g $y = 4\cos\frac{1}{2}x$ h $y = 3\cos 2x$

8.8 You can use the quotient rule, together with the results obtained for the derivatives of $\sin x$ and $\cos x$, to differentiate $\tan x$.

Let $y = \tan x$.

Then $y = \dfrac{\sin x}{\cos x}$, which is a quotient.

> This is the definition for $\tan x$.

Use the quotient rule

$$\frac{dy}{dx} = \frac{v\dfrac{du}{dx} - u\dfrac{dv}{dx}}{v^2}$$

with $u = \sin x$ and $v = \cos x$.

Then $\dfrac{dy}{dx} = \dfrac{\cos x \cos x - \sin x(-\sin x)}{\cos^2 x}$

$$= \frac{\cos^2 x + \sin^2 x}{\cos^2 x}$$

> Using the result that $\sin^2 x + \cos^2 x = 1$.

$$= \frac{1}{\cos^2 x}$$

$$= \sec^2 x$$

> As $\sec x = \dfrac{1}{\cos x}$

■ So if x is measured in radians

- $y = \tan x$ implies that $\dfrac{dy}{dx} = \sec^2 x$

> You should learn these key points.

■ Also by the chain rule, if $y = \tan f(x)$ then $\dfrac{dy}{dx} = f'(x)\sec^2 f(x)$

Example 13

Differentiate   **a** $y = x\tan 2x$ **b** $y = \tan^4 x$

a $y = x\tan 2x$

$\dfrac{dy}{dx} = x2\sec^2 2x + \tan 2x$

$= 2x\sec^2 2x + \tan 2x$

> This is a product.
> Use $u = x$ and $v = \tan 2x$, together with the product formula.

b $y = \tan^4 x = (\tan x)^4$

$\dfrac{dy}{dx} = 4(\tan x)^3(\sec^2 x)$

$= 4\tan^3 x \sec^2 x$

> Use the chain rule with $u = \tan x$.

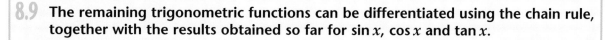

Exercise 8H

1 Differentiate:

   a $y = \tan 3x$ **b** $y = 4\tan^3 x$ **c** $y = \tan(x - 1)$ **d** $y = x^2 \tan\frac{1}{2}x + \tan(x - \frac{1}{2})$

8.9 The remaining trigonometric functions can be differentiated using the chain rule, together with the results obtained so far for $\sin x$, $\cos x$ and $\tan x$.

Let $y = \operatorname{cosec} x$.

Then $y = \dfrac{1}{\sin x} = (\sin x)^{-1}$

So $\dfrac{dy}{dx} = -(\sin x)^{-2}(\cos x)$

$\qquad = -\dfrac{\cos x}{\sin^2 x}$ •————————— This equals $-\dfrac{1}{\sin x} \times \dfrac{\cos x}{\sin x}$

$\qquad = -\operatorname{cosec} x \cot x$

- $y = \operatorname{cosec} x$ implies that $\dfrac{dy}{dx} = -\operatorname{cosec} x \cot x$

- Also by the chain rule, if $y = \operatorname{cosec} f(x)$ then $\dfrac{dy}{dx} = -f'(x)\operatorname{cosec} f(x)\cot f(x)$

Let $y = \sec x$.

Then $y = \dfrac{1}{\cos x} = (\cos x)^{-1}$

So $\dfrac{dy}{dx} = -(\cos x)^{-2}(-\sin x)$

$\qquad = \dfrac{\sin x}{\cos^2 x}$ •————————— Note that $\dfrac{\sin x}{\cos^2 x} = \dfrac{1}{\cos x} \times \dfrac{\sin x}{\cos x}$

$\qquad = \sec x \tan x$

- $y = \sec x$ implies that $\dfrac{dy}{dx} = \sec x \tan x$

- Also by the chain rule, if $y = \sec f(x)$ then $\dfrac{dy}{dx} = f'(x)\sec f(x)\tan f(x)$

Let $y = \cot x$.

Then $y = \dfrac{1}{\tan x} = (\tan x)^{-1}$

So $\dfrac{dy}{dx} = -(\tan x)^{-2}(\sec^2 x)$

$\qquad = \dfrac{\sec^2 x}{\tan^2 x}$ •————————— $\dfrac{\sec^2 x}{\tan^2 x} = \dfrac{1}{\cos^2 x} \times \dfrac{\cos^2 x}{\sin^2 x} = \dfrac{1}{\sin^2 x}$

$\qquad = -\operatorname{cosec}^2 x$

- $y = \cot x$ implies that $\dfrac{dy}{dx} = -\operatorname{cosec}^2 x$

- Also by the chain rule, if $y = \cot f(x)$ then $\dfrac{dy}{dx} = -f'(x)\operatorname{cosec}^2 f(x)$

Collecting all these results together,

- $y = \sin x \Rightarrow \dfrac{dy}{dx} = \cos x$

- $y = \cos x \Rightarrow \dfrac{dy}{dx} = -\sin x$

- $y = \tan x \Rightarrow \dfrac{dy}{dx} = \sec^2 x$

- $y = \operatorname{cosec} x \Rightarrow \dfrac{dy}{dx} = -\operatorname{cosec} x \cot x$

- $y = \sec x \Rightarrow \dfrac{dy}{dx} = \sec x \tan x$

- $y = \cot x \Rightarrow \dfrac{dy}{dx} = -\operatorname{cosec}^2 x$

> These results, obtained so far, should all be learned. They can be used together with the chain rule and the product and quotient rules to enable you to differentiate a wide range of functions.

Example 14

Differentiate **a** $y = \dfrac{\operatorname{cosec} 2x}{x^2}$ **b** $y = \sec^3 x$

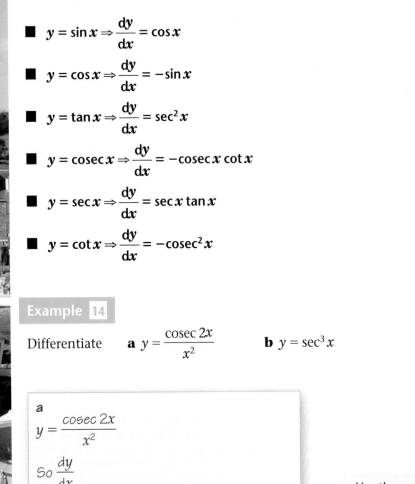

a
$$y = \frac{\operatorname{cosec} 2x}{x^2}$$

So $\dfrac{dy}{dx}$

$$= \frac{x^2(-2\operatorname{cosec} 2x \cot 2x) - \operatorname{cosec} 2x \times 2x}{x^4}$$

$$= \frac{-2\operatorname{cosec} 2x(x \cot 2x + 1)}{x^3}$$

Use the quotient rule with $u = \operatorname{cosec} 2x$ and $v = x^2$.

b $y = \sec^3 x = (\sec x)^3$

Use the chain rule with $u = \sec x$.

$$\frac{dy}{dx} = 3(\sec x)^2(\sec x \tan x)$$

$$= 3\sec^3 x \tan x$$

Exercise 8I

1 Differentiate

   a $\cot 4x$ **b** $\sec 5x$ **c** $\operatorname{cosec} 4x$ **d** $\sec^2 3x$

   e $x \cot 3x$ **f** $\dfrac{\sec^2 x}{x}$ **g** $\operatorname{cosec}^3 2x$ **h** $\cot^2(2x - 1)$

8.10 You are now able to differentiate functions that are formed from a combination of trigonometric, exponential, logarithmic and polynomial functions.

Example 15

Differentiate   **a** $y = e^x \sin x$ **b** $y = \dfrac{\ln x}{\sin x}$

a $y = e^x \sin x$

$\dfrac{dy}{dx} = e^x \cos x + e^x \sin x$ ⟵ Use the product rule with $u = e^x$ and $v = \sin x$.

b $y = \dfrac{\ln x}{\sin x}$

$\dfrac{dy}{dx} = \dfrac{\sin x \times \dfrac{1}{x} - \ln x \times \cos x}{\sin^2 x}$ ⟵ Use the quotient rule with $u = \ln x$ and $v = \sin x$.

$= \dfrac{\sin x - x \cos x \ln x}{x \sin^2 x}$

Exercise 8J

1 Find the function $f'(x)$ where $f(x)$ is

a $\sin 3x$ **b** $\cos 4x$ **c** $\tan 5x$ **d** $\sec 7x$

e $\operatorname{cosec} 2x$ **f** $\cot 3x$ **g** $\sin \dfrac{2x}{5}$ **h** $\cos \dfrac{3x}{7}$

i $\tan \dfrac{2x}{5}$ **j** $\operatorname{cosec} \dfrac{x}{2}$ **k** $\cot \tfrac{1}{3}x$ **l** $\sec \dfrac{3x}{2}$

2 Find the function $f'(x)$ where $f(x)$ is

a $\sin^2 x$ $2(\sin x)(\cos x)$ $\sin 2x$ **b** $\cos^3 x$ $3(\cos x)^2 (-\sin x)$ **c** $\tan^4 x$ $4(\tan)^3 \times \sec^2 x$ $\dfrac{4\sin^3}{\cos^5}$ **d** $(\sec x)^{\frac{1}{2}}$ $\tfrac{1}{2}\times(\cos)^{\frac{1}{2}}\times \sec x \tan$

e $\sqrt{\cot x}$ **f** $\operatorname{cosec}^2 x$ **g** $\sin^3 x$ **h** $\cos^4 x$

i $\tan^2 x$ **j** $\sec^3 x$ $3x(\sec)^2 \times \sec x \tan = 3\sec^3 x \tan$ **k** $\cot^3 x$ **l** $\operatorname{cosec}^4 x$ $4(\cos ec)^3 x$ $-\operatorname{cosec} \cot$ $-4\operatorname{cosec}^4 x \cot$

3 Find the function $f'(x)$ where $f(x)$ is

a $x \cos x$ **b** $x^2 \sec 3x$ **c** $\dfrac{\tan 2x}{x}$ **d** $\sin^3 x \cos x$ $3\sin^2 x \cos^2 + -\sin^4$

e $\dfrac{x^2}{\tan x}$ $\dfrac{\tan x 2x - \sec^2 x x^2}{\tan^2}$ **f** $\dfrac{1 + \sin x}{\cos x}$ **g** $e^{2x} \cos x$ $e^{2x}x - \sin + \cos x 2 e^{2x}$ **h** $e^x \sec 3x$

i $\dfrac{\sin 3x}{e^x}$ **j** $e^x \sin^2 x$ **k** $\dfrac{\ln x}{\tan x}$ $\dfrac{\tan x \times \frac{1}{x} - \ln x \times \sec^2}{\tan^2}$ **l** $\dfrac{e^{\sin x}}{\cos x}$ $\dfrac{e^{\sin}\cos^2 + \sin x \, e^{\sin x}}{\cos^2}$

1 Differentiate with respect to x:

    a $\ln x^2$ **b** $x^2 \sin 3x$ **E**

2 Given that

$$f(x) \equiv 3 - \frac{x^2}{4} + \ln\frac{x}{2}, \quad x > 0$$

    find $f'(x)$.   **E**

3 Given that $2y = x - \sin x \cos x$, show that $\dfrac{dy}{dx} = \sin^2 x$.   **E**

4 Differentiate, with respect to x,

    a $\dfrac{\sin x}{x}, \quad x > 0$ **b** $\ln\dfrac{1}{x^2 + 9}$ **E**

5 Use the derivatives of $\sin x$ and $\cos x$ to prove that the derivative of $\tan x$ is $\sec^2 x$.   **E**

6 $f(x) = \dfrac{x}{x^2 + 2}, \quad x \in \mathbb{R}$

    Find the set of values of x for which $f'(x) < 0$.   **E**

7 The function f is defined for positive real values of x by

$$f(x) = 12 \ln x - x^{\frac{3}{2}}$$

    Write down the set of values of x for which $f(x)$ is an increasing function of x.   **E**

8 Given that $y = \cos 2x + \sin x$, $0 < x < 2\pi$, and x is in radians, find, to 2 decimal places, the values of x for which $\dfrac{dy}{dx} = 0$.   **E**

9 The maximum point on the curve with equation $y = x\sqrt{\sin x}$, $0 < x < \pi$, is the point A. Show that the x-coordinate of point A satisfies the equation $2\tan x + x = 0$.   **E**

10 $f(x) = e^{0.5x} - x^2, \quad x \in \mathbb{R}$

    a Find $f'(x)$.

    b By evaluating $f'(6)$ and $f'(7)$, show that the curve with equation $y = f(x)$ has a stationary point at $x = p$, where $6 < p < 7$.   **E**

11 $f(x) \equiv e^{2x} \sin 2x, \quad 0 \le x \le \pi$

    a Use calculus to find the coordinates of the turning points on the graph of $y = f(x)$.

    b Show that $f''(x) = 8e^{2x} \cos 2x$.

    c Hence, or otherwise, determine which turning point is a maximum and which is a minimum.   **E**

12 The curve **C** has equation $y = 2e^x + 3x^2 + 2$. The point A with coordinates $(0, 4)$ lies on **C**. Find the equation of the tangent to **C** at A.   **E**

13 The curve **C** has equation $y = f(x)$, where

$$f(x) = 3\ln x + \frac{1}{x}, \quad x > 0$$

The point P is a stationary point on **C**.

a Calculate the x-coordinate of P.

The point Q on **C** has x-coordinate 1.

b Find an equation for the normal to **C** at Q. **E**

14 Differentiate $e^{2x}\cos x$ with respect to x.
The curve **C** has equation $y = e^{2x}\cos x$.

a Show that the turning points on **C** occur when $\tan x = 2$.

b Find an equation of the tangent to **C** at the point where $x = 0$. **E**

15 Given that $x = y^2 \ln y$, $y > 0$,

a find $\dfrac{dx}{dy}$

b use your answer to part **a** to find in terms of e, the value of $\dfrac{dy}{dx}$ at $y = e$. **E**

16

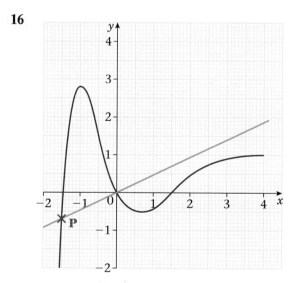

The figure shows part of the curve **C** with equation $y = f(x)$, where $f(x) = (x^3 - 2x)e^{-x}$

a Find $f'(x)$.

The normal to **C** at the origin O intersects **C** at a point **P**, as shown in the figure.

b Show that the x-coordinate of **P** is the solution of the equation

$$2x^2 = e^x + 4.$$ **E**

17

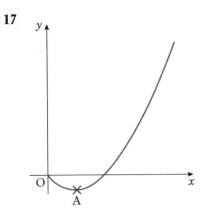

The diagram shows part of the curve with equation $y = f(x)$ where

$$f(x) = x(1 + x) \ln x \quad \{x > 0\}$$

The point A is the minimum point of the curve.

a Find $f'(x)$.

b Hence show that the x-coordinate of A is the solution of the equation $x = g(x)$, where

$$g(x) = e^{-\frac{1 + x}{1 + 2x}}$$

E

Summary of key points

You should learn all of these results.

1 You can use the chain rule to differentiate a function of a function:

- if $y = [f(x)]^n$ then $\dfrac{dy}{dx} = n[f(x)]^{n-1} f'(x)$

- if $y = f[g(x)]$ then $\dfrac{dy}{dx} = f'[g(x)]g'(x)$

2 Another form of the **chain rule** states that $\dfrac{dy}{dx} = \dfrac{dy}{du} \times \dfrac{du}{dx}$ where y is a function of u, and u is a function of x.

3 A particular case of the chain rule is the result $\dfrac{dy}{dx} = \dfrac{1}{\left(\dfrac{dx}{dy}\right)}$

4 You can use the product rule when two functions $u(x)$ and $v(x)$ are multiplied together.

- If $y = uv$ then $\dfrac{dy}{dx} = u\dfrac{dv}{dx} + v\dfrac{du}{dx}$

5 You can use the quotient rule when one function $u(x)$ is divided by another function $v(x)$, to form a rational function.

- If $y = \dfrac{u}{v}$ then $\dfrac{dy}{dx} = \dfrac{v\dfrac{du}{dx} - u\dfrac{dv}{dx}}{v^2}$

6 If $y = e^x$ then $\dfrac{dy}{dx} = e^x$ also and if $y = e^{f(x)}$ then $\dfrac{dy}{dx} = f'(x)e^{f(x)}$

7 If $y = \ln x$ then $\dfrac{dy}{dx} = \dfrac{1}{x}$ and if $y = \ln[f(x)]$ then $\dfrac{dy}{dx} = \dfrac{f'(x)}{f(x)}$

8 If $y = \sin x$ then $\dfrac{dy}{dx} = \cos x$.

9 If $y = \cos x$ then $\dfrac{dy}{dx} = -\sin x$.

10 If $y = \tan x$ then $\dfrac{dy}{dx} = \sec^2 x$.

11 If $y = \operatorname{cosec} x$ then $\dfrac{dy}{dx} = -\operatorname{cosec} x \cot x$.

12 If $y = \sec x$ then $\dfrac{dy}{dx} = \sec x \tan x$.

13 If $y = \cot x$ then $\dfrac{dy}{dx} = -\operatorname{cosec}^2 x$.

The chain rule can be used with each of these functions to obtain further results. (See the examples in the section.)

Review Exercise

1 a On the same set of axes sketch the graphs of $y = |2x - 1|$ and $y = |x - k|$, $k > 1$.

b Find, in terms of k, the values of x for which $|2x - 1| = |x - k|$.

2 a Sketch the graph of $y = |3x + 2| - 4$, showing the coordinates of the points of intersection of the graph with the axes.

b Find the values of x for which $|3x + 2| = 4 + x$.

3

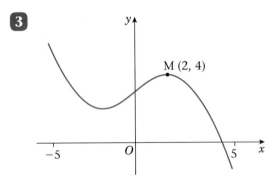

The figure shows the graph of $y = f(x)$, $-5 \leqslant x \leqslant 5$.

The point M (2, 4) is the maximum turning point of the graph.

Sketch, on separate diagrams, the graphs of

a $y = f(x) + 3$ **b** $y = |f(x)|$

c $y = f(|x|)$.

Show on each graph the coordinates of any maximum turning points. **E**

4

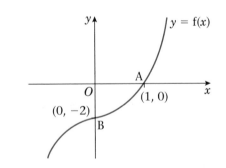

The diagram shows a sketch of the graph of the increasing function f.

The curve crosses the x-axis at the point A(1, 0) and the y-axis at the point B(0, −2)

On separate diagrams, sketch the graph of:

a $y = f^{-1}(x)$ **b** $y = f(|x|)$

c $y = f(2x) + 1$ **d** $y = 3f(x - 1)$.

In each case, show the images of the points A and B.

5 For the positive constant k, where $k > 1$, the functions f and g are defined by

$$f: \quad x \to \ln(x + k), x > -k,$$

$$g : x \to |2x - k|, \quad x \in \mathbb{R}$$

a Sketch, on the same set of axes, the graphs of f and g. Give the coordinates of points where the graphs meet the axes.

b Write down the range of f.

c Find, in terms of k, $\mathrm{fg}\!\left(\dfrac{k}{4}\right)$.

The curve C has equation $y = \mathrm{f}(x)$. The tangent to C at the point with x-coordinate 3 is parallel to the line with equation $9y = 2x + 1$.

d Find the value of k. **E**

6

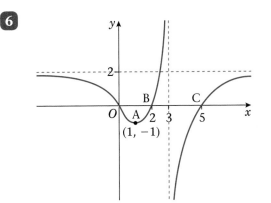

The diagram shows a sketch of the graph of $y = \mathrm{f}(x)$.

The curve has a minimum at the point $A(1, -1)$, passes through x-axis at the origin, and the points $B(2, 0)$ and $C(5, 0)$; the asymptotes have equations $x = 3$ and $y = 2$.

a Sketch, on separate axes, the graph of

i $y = |\mathrm{f}(x)|$
ii $y = -\mathrm{f}(x + 1)$
iii $y = \mathrm{f}(-2x)$

in each case, showing the images of the points A, B and C.

b State the number of solutions to the equation

i $3|\mathrm{f}(x)| = 2$ **ii** $2|\mathrm{f}(x)| = 3$.

7

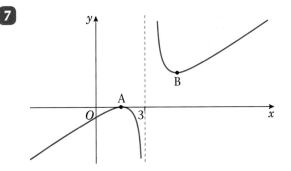

The diagram shows part of the curve C with equation $y = \mathrm{f}(x)$ where

$$\mathrm{f}(x) = \frac{(x - 1)^2}{(x - 3)}.$$

The points A and B are the stationary points of C.

The line $x = 3$ is a vertical asymptote to C.

a Using calculus, find the coordinates of A and B.

b Sketch the curve C^*, with equation $y = \mathrm{f}(-x) + 2$, showing the coordinates of the images of A and B.

c State the equation of the vertical asymptote to C^*.

8 **a** On the same set of axes, in the interval $-\pi < \theta < \pi$, sketch the graphs of

i $y = \cot \theta$,   **ii** $y = 3 \sin 2\theta$.

b Solve, in the interval $-\pi < \theta < \pi$, the equation

$$\cot \theta = 3 \sin 2\theta.$$

giving your answers, in radians, to 3 significant figures where appropriate.

9

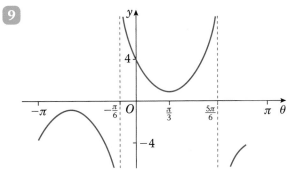

The diagram shows, in the interval $-\pi \leqslant \theta \leqslant \pi$, the graph of $y = k \sec(\theta - \alpha)$.

The curve crosses the y-axis at the point $(0, 4)$, and the θ-coordinate of its minimum point is $\dfrac{\pi}{3}$.

a State, as a multiple of π, the value of α.

b Find the value of k.

c Find the exact values of θ at the points where the graph crosses the line $y = -2\sqrt{2}$.

d Show that the gradient at the point on the curve with θ-coordinate $\dfrac{7\pi}{12}$ is $2\sqrt{2}$.

10 a Given that $\sin^2\theta + \cos^2\theta \equiv 1$, show that $1 + \tan^2\theta \equiv \sec^2\theta$.

b Solve, for $0 \le \theta < 360°$, the equation
$$2\tan^2\theta + \sec\theta = 1,$$
giving your answers to 1 decimal place.

(E)

11 a Prove that $\sec^4\theta - \tan^4\theta = 1 + 2\tan^2\theta$.

b Find all the values of x, in the interval $0 \le x \le 360°$, for which
$$\sec^4 2x = \tan 2x(3 + \tan^3 2x).$$
Give your answers correct to 1 decimal place, where appropriate.

12 a Prove that
$$\cot\theta - \tan\theta = 2\cot 2\theta, \quad \theta \ne \dfrac{n\pi}{2}.$$

b Solve, for $-\pi < \theta < \pi$, the equation
$$\cot\theta - \tan\theta = 5,$$
giving your answers to 3 significant figures.

13 a Solve, in the interval $0 \le \theta \le 2\pi$
$$\sec\theta + 2 = \cos\theta + \tan\theta\,(3 + \sin\theta),$$
giving your answers to 3 significant figures.

b Solve, in the interval $0 \le x \le 360°$,
$$\cot^2 x = \operatorname{cosec} x(2 - \operatorname{cosec} x),$$
giving your answers to 1 decimal place.

14 Given that
$$y = \arcsin x, \ -1 \le x \le 1 \text{ and } -\dfrac{\pi}{2} \le y \le \dfrac{\pi}{2},$$

a express $\arccos x$ in terms of y.

b Hence find, in terms of π the value of $\arcsin x + \arccos x$.

Given that
$$y = \arccos x, \ -1 \le x \le 1 \text{ and } 0 \le y \le \pi,$$

c sketch, on the same set of axes, the graphs of $y = \arcsin x$ and $y = \arccos x$, making it clear which is which.

d Explain how your sketches can be used to evaluate $\arcsin x + \arccos x$.

15 a By writing $\cos 3\theta$ as $\cos(2\theta + \theta)$, show that
$$\cos 3\theta = 4\cos^3\theta - 3\cos\theta.$$

b Given that $\cos\theta = \dfrac{\sqrt{2}}{3}$, find the exact value of $\sec 3\theta$.

16 Given that $\sin(x + 30°) = 2\sin(x - 60°)$,

a show that $\tan x = 8 + 5\sqrt{3}$.

b Hence express $\tan(x + 60°)$ in the form $a + b\sqrt{3}$.

17 a Given that $\cos A = \dfrac{3}{4}$ where $270° < A < 360°$, find the exact value of $\sin 2A$.

b **i** Show that
$$\cos\left(2x + \dfrac{\pi}{3}\right) + \cos\left(2x - \dfrac{\pi}{3}\right) = \cos 2x$$
Given that
$$y = 3\sin^2 x + \cos\left(2x + \dfrac{\pi}{3}\right) + \cos\left(2x - \dfrac{\pi}{3}\right),$$

ii show that $\dfrac{dy}{dx} = \sin 2x$ **(E)**

18 Solve, in the interval $-180° \le x < 180°$, the equations

a $\cos 2x + \sin x = 1$

b $\sin x\,(\cos x + \operatorname{cosec} x) = 2\cos^2 x$, giving your answers to 1 decimal place.

19 a Prove that
$$\dfrac{\sin\theta}{\cos\theta} + \dfrac{\cos\theta}{\sin\theta} = 2\operatorname{cosec} 2\theta, \quad \theta \ne 90n°.$$

b Sketch the graph of $y = 2\operatorname{cosec} 2\theta$ for $0° < \theta < 360°$.

c Solve, for $0° < \theta < 360°$, the equation

$$\frac{\sin \theta}{\cos \theta} + \frac{\cos \theta}{\sin \theta} = 3,$$

giving your answers to 1 decimal place.

E

20 a Express $3 \sin x + 2 \cos x$ in the form $R \sin (x + \alpha)$, where $R > 0$ and $0 < \alpha < \frac{\pi}{2}$.

b Hence find the greatest value of $(3 \sin x + 2 \cos x)^4$.

c Solve, for $0 < x < 2\pi$, the equation

$$3 \sin x + 2 \cos x = 1,$$

giving your answers to 3 decimal places.

E

21 The point P lies on the curve with equation $y = \ln\left(\frac{1}{3}x\right)$.

The x-coordinate of P is 3.

Find an equation of the normal to the curve at the point P in the form $y = ax + b$, where a and b are constants.

E

22 a Differentiate with respect to x

  i $3 \sin^2 x + \sec 2x$,

  ii $\{x + \ln(2x)\}^3$.

Given that $y = \dfrac{5x^2 - 10x + 9}{(x - 1)^2}$, $x \neq -1$,

b show that $\dfrac{dy}{dx} = -\dfrac{8}{(x - 1)^3}$

E

23 Given that $y = \ln(1 + e^x)$,

a show that when $x = -\ln 3$, $\dfrac{dy}{dx} = \dfrac{1}{4}$

b find the exact value of x for which

$$e^y \frac{dy}{dx} = 6.$$

24 a Differentiate with respect to x

  i $x^2 e^{3x + 2}$,

  ii $\dfrac{\cos(2x^3)}{3x}$.

b Given that $x = 4 \sin (2y + 6)$, find $\dfrac{dy}{dx}$ in terms of x.

E

25 Given that $x = y^2 e^{\sqrt{y}}$,

a find, in terms of y, $\dfrac{dx}{dy}$

b show that when $y = 4$, $\dfrac{dy}{dx} = \dfrac{e^{-2}}{12}$.

26 a Given that $y = \sqrt{(1 + x^2)}$, show that $\dfrac{dy}{dx} = \dfrac{\sqrt{3}}{2}$ when $x = \sqrt{3}$.

b Given that $y = \ln\{x + \sqrt{(1 + x^2)}\}$, show that $\dfrac{dy}{dx} = \dfrac{1}{\sqrt{(1 + x^2)}}$.

27 Given that $f(x) = x^2 e^{-x}$,

a find $f'(x)$, using the product rule for differentiation

b show that $f''(x) = (x^2 - 4x + 2)e^{-x}$.

A curve C has equation $y = f(x)$.

c Find the coordinates of the turning points of C.

d Determine the nature of each turning point of the curve C.

28 a Express $(\sin 2x + \sqrt{3} \cos 2x)$ in the form $R \sin(2x + k\pi)$, where $R > 0$ and $0 < k < \frac{1}{2}$.

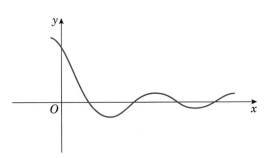

The diagram shows part of the curve with equation

$$y = e^{-2\sqrt{2}x}(\sin 2x + \sqrt{3} \cos 2x).$$

b Show that the x-coordinates of the turning points of the curve satisfy the equation

$$\tan\left(2x + \frac{\pi}{3}\right) = \frac{1}{\sqrt{2}}.$$

29 The curve C has equation $y = x^2\sqrt{\cos x}$. The point P on C has x-coordinate $\dfrac{\pi}{3}$.

a Show that the y-coordinate of P is $\dfrac{\sqrt{2}\,\pi^2}{18}$.

b Show that the gradient of C at P is 0.809, to 3 significant figures.

In the interval $0 < x < \dfrac{\pi}{2}$, C has a maximum at the point A.

c Show that the x-coordinate, k, of A satisfies the equation $x \tan x = 4$.

The iterative formula
$$x_{n+1} = \tan^{-1}\!\left(\frac{4}{x_n}\right), \quad x_0 = 1.25,$$
is used to find an approximation for k.

d Find the value of k, correct to 4 decimal places.

30

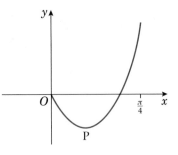

The figure shows part of the curve with equation
$$y = (2x - 1)\tan 2x, \quad 0 \le x < \frac{\pi}{4}.$$

The curve has a minimum at the point P. The x-coordinate of P is k.

a Show that k satisfies the equation
$$4k + \sin 4k - 2 = 0.$$

The iterative formula
$$x_{n+1} = \tfrac{1}{4}(2 - \sin 4x_n), \quad x_0 = 0.3,$$
is used to find an approximate value for k.

b Calculate the values of x_1, x_2, x_3 and x_4, giving your answers to 4 decimal places.

c Show that $k = 0.277$, correct to 3 significant figures. **E**

Practice paper

(Marks are shown in brackets.)

1 The curve C, with equation $y = x^2 \ln x$, $x > 0$, has a stationary point P. Find, in terms of e, the coordinates of P. (7)

2 $f(x) = e^{2x-1}$, $x \geqslant 0$

The curve C with equation $y = f(x)$ meets the y-axis at P.
The tangent to C at P crosses the x-axis at Q.

a Find, to 3 decimal places, the area of triangle POQ, where O is the origin. (5)

The line $y = 2$ intersects C at the point R.

b Find the exact value of the x-coordinate of R. (3)

3 $f(x) = \dfrac{3x}{x+1} - \dfrac{x+7}{x^2-1}$, $x > 1$

a Show that $f(x) = 3 - \dfrac{4}{x-1}$, $x > 1$. (5)

b Find $f^{-1}(x)$. (4)

c Write down the domain of $f^{-1}(x)$. (1)

4 a Sketch, on the same set of axes, for $x > 0$, the graphs of

$$y = -1 + \ln 3x \quad \text{and} \quad y = \frac{1}{x}$$ (2)

The curves intersect at the point P whose x-coordinate is p.
Show that

b p satisfies the equation

$$p \ln 3p - p - 1 = 0$$ (1)

c $1 < p < 2$ (2)

The iterative formula

$$x_{n+1} = \tfrac{1}{3}e^{\left(1 + \frac{1}{x_n}\right)}, \quad x_0 = 2$$

is used to find an approximation for p.

d Write down the values of x_1, x_2, x_3 and x_4 giving your answers to 4 significant figures. (3)

e Prove that $p = 1.66$ correct to 3 significant figures. (2)

5 The curve C_1 has equation

$$y = \cos 2x - 2 \sin^2 x$$

The curve C_2 has equation

$$y = \sin 2x$$

a Show that the x-coordinates of the points of intersection of C_1 and C_2 satisfy the equation

$$2\cos 2x - \sin 2x = 1 \qquad (3)$$

b Express $2\cos 2x - \sin 2x$ in the form $R\cos(2x + \alpha)$, where $R > 0$ and $0 < \alpha < \dfrac{\pi}{2}$, giving the exact value of R and giving α in radians to 3 decimal places. $\qquad (4)$

c Find the x-coordinates of the points of intersection of C_1 and C_2 in the interval $0 \leqslant x < \pi$, giving your answers in radians to 2 decimal places. $\qquad (5)$

6 a Given that $y = \ln \sec x$, $-\dfrac{\pi}{2} < x \leqslant 0$, use the substitution $u = \sec x$, or otherwise, to show that $\dfrac{dy}{dx} = \tan x$. $\qquad (3)$

The curve C has equation $y = \tan x + \ln \sec x$, $-\dfrac{\pi}{2} < x \leqslant 0$.

At the point P on C, whose x-coordinate is p, the gradient is 3.

b Show that $\tan p = -2$. $\qquad (6)$

c Find the exact value of $\sec p$, showing your working clearly. $\qquad (2)$

d Find the y-coordinate of P, in the form $a + k \ln b$, where a, k and b are rational numbers. $\qquad (2)$

7 The diagram shows a sketch of part of the curve with equation $y = f(x)$. The curve has no further turning points.

On separate diagrams show a sketch of the curve with equation

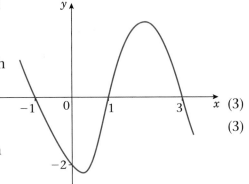

a $y = 2f(-x)$ $\qquad (3)$

b $y = |f(2x)|$ $\qquad (3)$

In each case show the coordinates of points in which the curve meets the coordinate axes.

The function g is given by

$$g: x \to |x + 1| - k, \quad x \in \mathbb{R}, \ k > 1$$

c Sketch the graph of g, showing, in terms of k, the y-coordinate of the point of intersection of the graph with the y-axis. $\qquad (3)$

Find, in terms of k,

d the range of $g(x)$ $\qquad (1)$

e $gf(0)$ $\qquad (2)$

f the solution of $g(x) = x$ $\qquad (3)$

Examination style paper

1 The point P lies on the curve with equation

$$y = [3x + \ln 2x]^2$$

The x-coordinate of P is 0.5.
Find an equation of the tangent to the curve at the point P. (6)

2 Express

$$\frac{3x^2 - 2x}{(x - 1)(3x - 2)} - \frac{2}{x^2 - 1}$$

as a single fraction in its simplest form. (7)

3 $\quad g(x) = 5 \sin x - 3 \cos x$

Given that $g(x) = R \sin(x - \alpha)$, where $R \geqslant 0$ and $0 < \alpha < 90°$

a find the value of R and the value of α. (4)

b i Write down the maximum value of $g(x)$. (1)

  ii Find, to the nearest degree, the smallest positive value of x for which the
maximum value occurs. (2)

4

          Figure 1

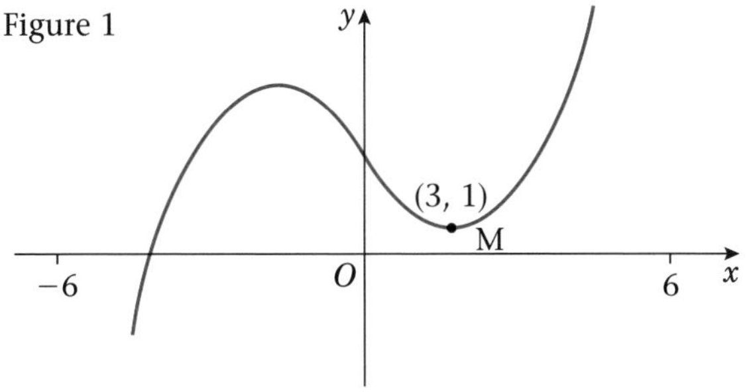

Figure 1 shows the graph of $y = f(x)$ $-6 \leqslant x \leqslant 6$. The point M(3, 1) is the minimum
turning point of the graph.

Sketch on separate diagrams the graphs of:

a $y = f(x) - 1$ (2)

b $y = |f(x + 3)|$ (3)

c $y = f(|x|)$ (3)

Show on each graph the coordinates of any minimum turning points.

5 $f(x) = x^5 - x^2 - 20$

   a Show that the equation $f(x) = 0$ can be written as

   $$x = \sqrt[3]{1 + \frac{20}{x^2}}$$

   The equation $f(x) = 0$ has a root α in the interval $[1, 2]$.   (3)

   b Use the iterative formula

   $$x_{n+1} = \sqrt[3]{1 + \frac{20}{x_n^2}}$$

   with $x_0 = 2$, to find, to 2 decimal places, the values of x_1, x_2, x_3 and x_4.   (4)

   c By choosing a suitable interval, prove that $\alpha = 1.88$ correct to 2 decimal places   (3)

6   The functions of f and g are defined by

   $$f: x \to \ln(3 - 2x) \qquad x \in \mathbb{R} \quad x < 1.5$$
   $$g: x \to e^{2x} + 1 \qquad x \in \mathbb{R}$$

   a Find $gf(-1)$.   (4)

   b Find $f^{-1}(x)$.   (3)

   c Find the exact value of x for which $f^{-1}(x) = g(x)$.   (5)

7   **a** Given that $\sin x = \frac{p}{q}$, $0 < x < \frac{\pi}{2}$ and $p > 0$ and $q > 0$,

   find $\text{cosec}2x$ in terms of p and q.   (4)

   b Solve for $0 < x < \pi$, giving your answers as multiples of π,

   $$2\cot^2 2x + 3\text{cosec}2x = 0$$ (8)

8   Differentiate with respect to x

   a $y = x^3 e^{2x-1}$ (4)

   b $y = \frac{\sin(x^2 + 1)}{2x}$ (4)

   c Given that $x = 3\tan(2y - 1)$, find $\frac{dy}{dx}$ in terms of x.   (5)

Formulae you need to remember

These are the formulae that you need to remember for your examinations. They will not be included in formulae booklets.

Trigonometry

$\cos^2 A + \sin^2 A \equiv 1$

$\sec^2 A \equiv 1 + \tan^2 A$

$\operatorname{cosec}^2 A \equiv 1 + \cot^2 A$

$\sin 2A \equiv 2 \sin A \cos A$

$\cos 2A \equiv \cos^2 A - \sin^2 A$

$\tan 2A \equiv \dfrac{2 \tan A}{1 - \tan^2 A}$

Differentiation

function	derivative
$\sin kx$	$k \cos kx$
$\cos kx$	$-k \sin kx$
e^{kx}	ke^{kx}
$\ln x$	$\dfrac{1}{x}$
$f(x) + g(x)$	$f'(x) + g'(x)$
$f(x)\,g(x)$	$f'(x)\,g(x) + f(x)\,g'(x)$
$f(g(x))$	$f'(g(x))\,g'(x)$

List of symbols and notation

The following notation will be used in all Edexcel mathematics examinations:

$\in$	is an element of
$\notin$	is not an element of
$\{x_1, x_2, \ldots\}$	the set with elements $x_1, x_2, \ldots$
$\{x: \ldots\}$	the set of all x such that $\ldots$
$n(A)$	the number of elements in set A
$\varnothing$	the empty set
ξ	the universal set
A'	the complement of the set A
$\mathbb{N}$	the set of natural numbers, $\{1, 2, 3, \ldots\}$
$\mathbb{Z}$	the set of integers, $\{0, \pm 1, \pm 2, \pm 3, \ldots\}$
$\mathbb{Z}^+$	the set of positive integers, $\{1, 2, 3, \ldots\}$
$\mathbb{Z}_n$	the set of integers modulo n, $\{1, 2, 3, \ldots, n-1\}$
$\mathbb{Q}$	the set of rational numbers, $\left\{\dfrac{p}{q}: p \in \mathbb{Z}_u, q \in \mathbb{Z}^+\right\}$
$\mathbb{Q}^+$	the set of positive rational numbers, $\{x \in \mathbb{Q}: x > 0\}$
$\mathbb{Q}_0^+$	the set of positive rational numbers and zero, $\{x \in \mathbb{Q}: x \geqslant 0\}$
$\mathbb{R}$	the set of real numbers
$\mathbb{R}^+$	the set of positive real numbers, $\{x \in \mathbb{R}: x > 0\}$
$\mathbb{R}_0^+$	the set of positive real numbers and zero, $\{x \in \mathbb{R}: x \geqslant 0\}$
$\mathbb{C}$	the set of complex numbers
(x, y)	the ordered pair x, y
$A \times B$	the Cartesian products of sets A and B, i.e. $A \times B = \{(a, b): a \in A, b \in B\}$
$\subseteq$	is a subset of
$\subset$	is a proper subset of
$\cup$	union
$\cap$	intersection
$[a, b]$	the closed interval, $\{x \in \mathbb{R}: a \leqslant x \leqslant b\}$
$[a, b), [a, b[$	the interval, $\{x \in \mathbb{R}: a \leqslant x < b\}$
$(a, b],]a, b]$	the interval, $\{x \in \mathbb{R}: a < x \leqslant b\}$
$(a, b),]a, b[$	the open interval, $\{x \in \mathbb{R}: a < x < b\}$
$y \, R \, x$	y is related to x by the relation R
$y \sim x$	y is equivalent to x, in the context of some equivalence relation
$=$	is equal to
$\neq$	is not equal to
$\equiv$	is identical to or is congruent to

$\approx$	is approximately equal to
$\equiv$	is isomorphic to
$\propto$	is proportional to
$<$	is less than
$\leqslant$, $\ngtr$	is less than or equal to, is not greater than
$>$	is greater than
$\geqslant$, $\nless$	is greater than or equal to, is not less than
∞	infinity
$p \wedge q$	p and q
$p \vee q$	p or q (or both)
$\sim p$	not p
$p \Rightarrow q$	p implies q (if p then q)
$p \Leftarrow q$	p is implied by q (if q then p)
$p \Leftrightarrow q$	p implies and is implied by q (p is equivalent to q)
$\exists$	there exists
$\forall$	for all
$a + b$	a plus b
$a - b$	a minus b
$a \times b$, ab, $a.b$	a multiplied by b
$a \div b$, $\dfrac{a}{b}$, a/b	a divided by b
$\displaystyle\sum_{i=1}^{n}$	$a_1 + a_2 + \ldots + a_n$
$\displaystyle\prod_{i=1}^{n}$	$a_1 \times a_2 \times \ldots \times a_n$
$\sqrt{a}$	the positive square root of a
$|a|$	the modulus of a
$n!$	n factorial
$\dbinom{n}{r}$	the binomial coefficient $\dfrac{n!}{r!(n-r)!}$ for $n \in \mathbb{Z}^+$
	$\dfrac{n(n-1) \ldots (n-r+1)}{r!}$ for $n \in \mathbb{Q}$
$f(x)$	the value of the function f at x
$f : A \rightarrow B$	f is a function under which each element of set A has an image in set B
$f : x \rightarrow y$	the function f maps the element x to the element y
f^{-1}	the inverse function of the function f
$g \circ f$, gf	the composite function of f and g which is defined by $(g \circ f)(x)$ or $gf(x) = g(f(x))$
$\displaystyle\lim_{x \to a} f(x)$	the limit of $f(x)$ as x tends to a
Δx, δx	an increment of x
$\dfrac{dy}{dx}$	the derivative of y with respect to x
$\dfrac{d^n y}{dx^n}$	the nth derivative of y with respect to x
$f'(x), f''(x), \ldots, f^{(n)}(x)$	the first, second, ..., nth derivatives of $f(x)$ with respect to x
$\displaystyle\int y \, dx$	the indefinite integral of y with respect to x

$\int_b^a y \, dx$	the definite integral of y with respect to x between the limits
$\dfrac{\partial V}{\partial x}$	the partial derivative of V with respect to x
$\dot{x}, \ddot{x}, \ldots$	the first, second, ... derivatives of x with respect to t
e	base of natural logarithms
e^x, exp x	exponential function of x
$\log_a x$	logarithm to the base a of x
ln x, $\log_e x$	natural logarithm of x
lg x, $\log_{10} x$	logarithm of x to base 10
sin, cos, tan, cosec, sec, cot	the circular functions
arcsin, arccos, arctan, arccosec, arcsec, arccot	the inverse circular functions
sinh, cosh, tanh, cosech, sech, coth	the hyperbolic functions
arsinh, arcosh, artanh, arcosech, arsech, arcoth	the inverse hyperbolic functions
i, j	square root of -1
z	a complex number, $z = x + iy$
Re z	the real part of z, Re $z = x$
Im z	the imaginary part of z, Im $z = y$
$|z|$	the modulus of z, $|z| = \sqrt{(x^2 + y^2)}$
arg z	the argument of z, arg $z = \theta$, $-\pi < \theta \leqslant \pi$
z^*	the complex conjugate of z, $x - iy$
$\mathbf{M}$	a matrix $\mathbf{M}$
$\mathbf{M}^{-1}$	the inverse of the matrix $\mathbf{M}$
$\mathbf{M}^{\mathrm{T}}$	the transpose of the matrix $\mathbf{M}$
det $\mathbf{M}$ or $|\mathbf{M}|$	the determinant of the square matrix $\mathbf{M}$
$\mathbf{a}$	the vector $\mathbf{a}$
$\overrightarrow{AB}$	the vector represented in magnitude and direction by the directed line segment AB
$\hat{a}$	a unit vector in the direction of $\mathbf{a}$
$\mathbf{i}, \mathbf{j}, \mathbf{k}$	unit vectors in the direction of the Cartesian coordinate axes
$|\mathbf{a}|$, a	the magnitude of $\mathbf{a}$
$|\overrightarrow{AB}|$	the magnitude of $\overrightarrow{AB}$
$\mathbf{a} . \mathbf{b}$	the scalar product of $\mathbf{a}$ and $\mathbf{b}$
$\mathbf{a} \times \mathbf{b}$	the vector product of $\mathbf{a}$ and $\mathbf{b}$

Answers

Exercise 1A

1 a 4 **b** $\frac{1}{3}$ **c** $\frac{x+4}{x+2}$ **d** $\frac{1}{4}$

e $\frac{2}{3}$ **f** $\frac{a+3}{a+6}$ **g** $\frac{1}{2}$ **h** $\frac{1}{4}$

i $\frac{x}{x+3}$ **j** $\frac{x}{x+3}$ **k** $\frac{x+1}{x+4}$ **l** $\frac{x^2}{x+2}$

m $\frac{x^2-4}{x^2+4}$ **n** $\frac{1}{x+3}$ **o** $\frac{x-3}{x-4}$ **p** $\frac{2(x-2)}{x+2}$

q $6(x-1)$ **r** $3(x+1)$

Exercise 1B

1 a $\frac{a^2}{cd}$ **b** a **c** $\frac{1}{2}$ **d** $\frac{1}{2}$

e $\frac{4}{x^2}$ **f** $\frac{r^5}{10}$ **g** $\frac{1}{x-2}$ **h** $\frac{a-3}{2(a+3)}$

i $\frac{x-3}{y}$ **j** $\frac{y+1}{y}$ **k** $\frac{x}{6}$ **l** 4

m $\frac{1}{x+5}$ **n** $\frac{3y-2}{2}$ **o** $\frac{2(x+y)^2}{(x-y)^2}$

Exercise 1C

1 a $\frac{q+p}{pq}$ **b** $\frac{a-b}{b}$ **c** $\frac{3}{2x}$

d $\frac{3-x}{x^2}$ **e** $\frac{7}{8x}$ **f** $\frac{x^2+y^2}{xy}$

g $-\frac{1}{(x+2)(x+1)}$ **h** $\frac{x-7}{(x+3)(x-2)}$ **i** $\frac{-x-5}{6}$

j $\frac{2x-4}{(x+4)^2}$ **k** $\frac{5x+3}{6(x+3)(x-1)}$

l $\frac{x+3}{(x+1)^2}$ **m** $\frac{x+4}{(x+1)(x+2)^2}$

n $\frac{-a-7}{(a+1)(a+3)^2}$ **o** $\frac{2+3y+3x}{(y+x)(y-x)}$

p $\frac{7x+8}{(x+2)(x+3)(x-4)}$ **q** $\frac{4x^2+17x+13}{(x+2)^2}$

Exercise 1D

1 a $x^2+3x+6+\frac{2}{x-1}$ **b** $2x^2-3x+5-\frac{10}{x+3}$

c x^2+2x+4 **d** $2+\frac{4x+7}{x^2-1}$

e $4x+1+\frac{-8x+3}{2x^2+2}$ **f** $4x-13+\frac{33x-27}{x^2+2x-1}$

g $x^2+2-\frac{6}{x^2+1}$ **h** x^3-x^2+x-1

i $2x^2+x+1+\frac{5x-4}{x^2+x-2}$

2 $A=3,\ B=-4,\ C=1,\ D=4,\ E=1$

Mixed exercise 1E

1 a $\frac{bc}{a}$ **b** $\frac{x+1}{4}$

c $2x$ **d** $\frac{x-1}{x}$

e $\frac{a+4}{a+8}$ **f** $\frac{b+5}{b+3}$

2 a $\frac{7x}{12}$ **b** $\frac{5}{2y}$

c $\frac{x+7}{6}$ **d** $x-4-\frac{10}{x-1}$

e $x^2-2x+11-\frac{23}{x+2}$ **f** $x^2-1+\frac{4}{x^2+1}$

3 $A=1,\ B=-4,\ C=3,\ D=8$
5 $B=1,\ C=3$
6 $A=2,\ B=-4,\ C=6,\ D=-11$

Exercise 2A

1 a

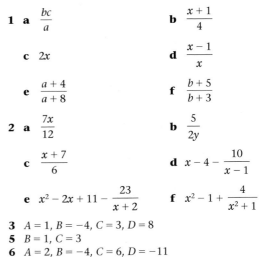
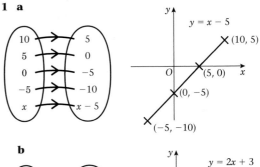
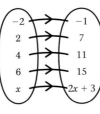
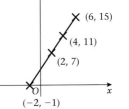
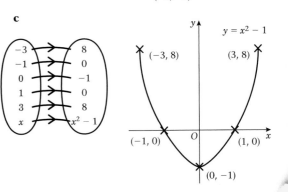

b

c

d

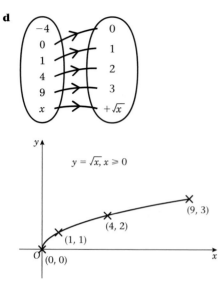

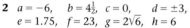

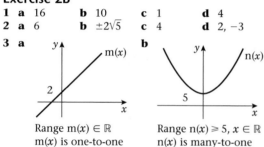

$y = \sqrt{x},\ x \geqslant 0$

(9, 3)

(4, 2)

(1, 1)

(0, 0)

2 $a = -6,\quad b = 4\frac{1}{2},\quad c = 0,\quad d = \pm 3,$
$e = 1.75,\quad f = 23,\quad g = 2\sqrt{6},\quad h = 6$

Exercise 2B

1 a 16   **b** 10   **c** 1   **d** 4
2 a 6   **b** $\pm 2\sqrt{5}$ **c** 4   **d** 2, -3

3 a

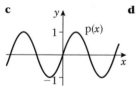

m(x)

2

Range $m(x) \in \mathbb{R}$
m(x) is one-to-one

b

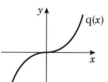

n(x)

5

Range $n(x) \geqslant 5,\ x \in \mathbb{R}$
n(x) is many-to-one

c

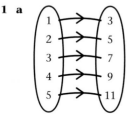

p(x)

1

-1

Range $-1 \leqslant p(x) \leqslant 1$
p(x) is many-to-one

d

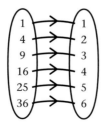

q(x)

Range $q(x) \in \mathbb{R}$
q(x) is one-to-one

4 a one-to-one function
b one-to-one function
c Not a function. Some x values get mapped to 2 y values.
d Not a function. Some x values get mapped to 2 y values.
e Not a function. One x value doesn't get mapped anywhere.
f many-to-one function

Exercise 2C

1 a

one-to-one function

b

one-to-one function

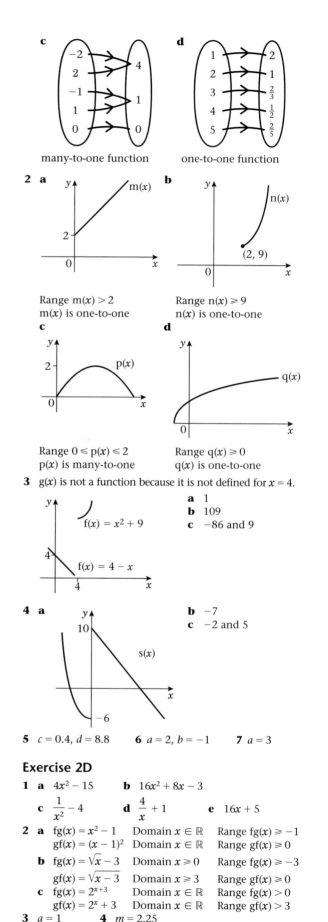

c

many-to-one function

d

one-to-one function

2 a

m(x)

2

Range $m(x) > 2$
m(x) is one-to-one

b

n(x)

(2, 9)

Range $n(x) \geqslant 9$
n(x) is one-to-one

c

p(x)

2

Range $0 \leqslant p(x) \leqslant 2$
p(x) is many-to-one

d

q(x)

Range $q(x) \geqslant 0$
q(x) is one-to-one

3 $g(x)$ is not a function because it is not defined for $x = 4$.

$f(x) = x^2 + 9$

$f(x) = 4 - x$

  a 1
  b 109
  c -86 and 9

4 a

s(x)

10

-6

  b -7
  c -2 and 5

5 $c = 0.4,\ d = 8.8$ **6** $a = 2,\ b = -1$ **7** $a = 3$

Exercise 2D

1 a $4x^2 - 15$ **b** $16x^2 + 8x - 3$
  c $\dfrac{1}{x^2} - 4$ **d** $\dfrac{4}{x} + 1$ **e** $16x + 5$

2 a $fg(x) = x^2 - 1$ Domain $x \in \mathbb{R}$ Range $fg(x) \geqslant -1$
    $gf(x) = (x - 1)^2$ Domain $x \in \mathbb{R}$ Range $gf(x) \geqslant 0$
  b $fg(x) = \sqrt{x} - 3$ Domain $x \geqslant 0$ Range $fg(x) \geqslant -3$
    $gf(x) = \sqrt{x - 3}$ Domain $x \geqslant 3$ Range $gf(x) \geqslant 0$
  c $fg(x) = 2^{x+3}$ Domain $x \in \mathbb{R}$ Range $fg(x) > 0$
    $gf(x) = 2^x + 3$ Domain $x \in \mathbb{R}$ Range $gf(x) > 3$

3 $a = 1$ **4** $m = 2.25$

5 a $l^2(x)$ **b** $ml(x)$ **c** $nm(x)$ **d** $ln(x)$
  e $pm(x)$ **f** $lm(x)$ **g** $p^3(x)$

8 $f^3(x) = \dfrac{x + 2}{2x + 3}$

Exercise 2E

1 a

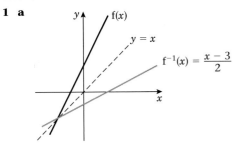

b

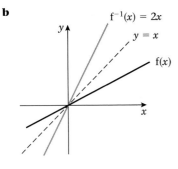

c

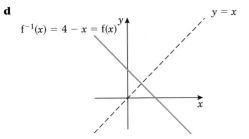

d

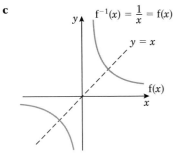

e

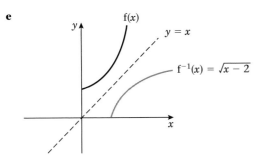

f

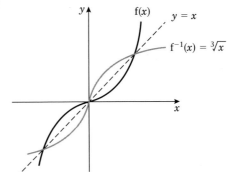

2 **c** and **d** are self inverses

3 $g(x) = 4 - x \; \{x \in \mathbb{R}, x > 0\}$
  whereas $g^{-1}(x) = 4 - x \; \{x \in \mathbb{R}, x < 4\}$

4 a

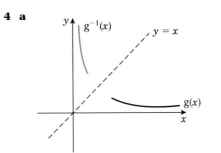

b

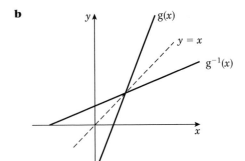

c

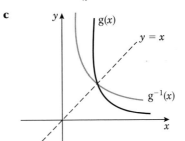

d

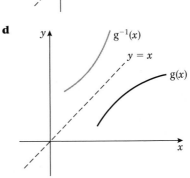

e

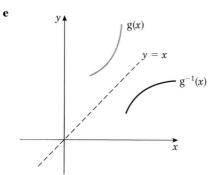

f

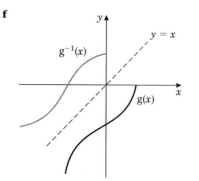

5 $a = -2$ $m^{-1}(x) = \sqrt{x - 5} - 2$ $\{x \in \mathbb{R}, x > 5\}$

6 $t^{-1}(x) = \sqrt{x + 4} + 3$ $\{x \in \mathbb{R}, x \geqslant 0\}$

7 a tends to infinity

  b 7

  c $h^{-1}(x) = \dfrac{2x + 1}{x - 2}$ $\{x \in \mathbb{R}, x \neq 2\}$

  d $2 + \sqrt{5}, 2 - \sqrt{5}$

8 a $f^{-1}(x) = -\sqrt{\dfrac{x + 3}{2}}$ $\{x \in \mathbb{R}, x > -3\}$

  b $a = -1$

Mixed exercise 2F

1 a not a function   **b** one-to-one function
  c many-to-one function **d** many-to-one function
  e not a function   **f** not a function

2 a $20, 28, \frac{1}{9}$ **b** $f(x) \geqslant -8, g(x) \in \mathbb{R}$

  c $g^{-1}(x) = \sqrt[3]{x - 1}$ $\{x \in \mathbb{R}\}$
  d $4(x^3 - 1)$ **e** $a = \frac{5}{3}$

3 a 8, 9   **b** -45 and $5\sqrt{2}$

4 a

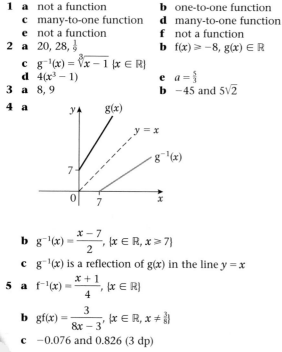

  b $g^{-1}(x) = \dfrac{x - 7}{2}, \{x \in \mathbb{R}, x \geqslant 7\}$

  c $g^{-1}(x)$ is a reflection of $g(x)$ in the line $y = x$

5 a $f^{-1}(x) = \dfrac{x + 1}{4}, \{x \in \mathbb{R}\}$

  b $gf(x) = \dfrac{3}{8x - 3}, \{x \in \mathbb{R}, x \neq \frac{3}{8}\}$

  c -0.076 and 0.826 (3 dp)

6 a   **b** $\frac{1}{2}$ and $1\frac{1}{2}$

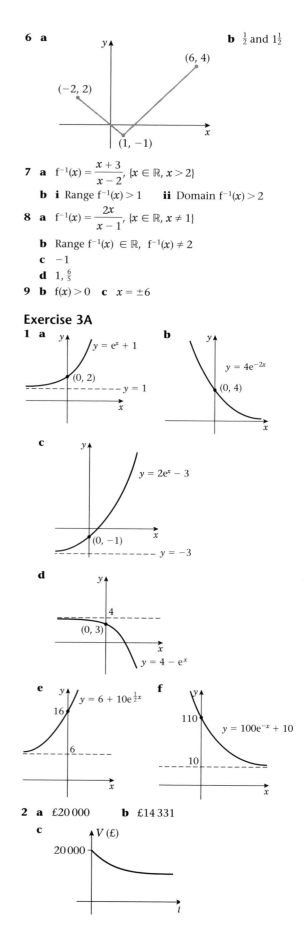

7 a $f^{-1}(x) = \dfrac{x + 3}{x - 2}, \{x \in \mathbb{R}, x > 2\}$

  b i Range $f^{-1}(x) > 1$ **ii** Domain $f^{-1}(x) > 2$

8 a $f^{-1}(x) = \dfrac{2x}{x - 1}, \{x \in \mathbb{R}, x \neq 1\}$

  b Range $f^{-1}(x) \in \mathbb{R}$, $f^{-1}(x) \neq 2$
  c -1
  d $1, \frac{6}{5}$

9 b $f(x) > 0$ **c** $x = \pm 6$

Exercise 3A

2 a £20 000   **b** £14 331

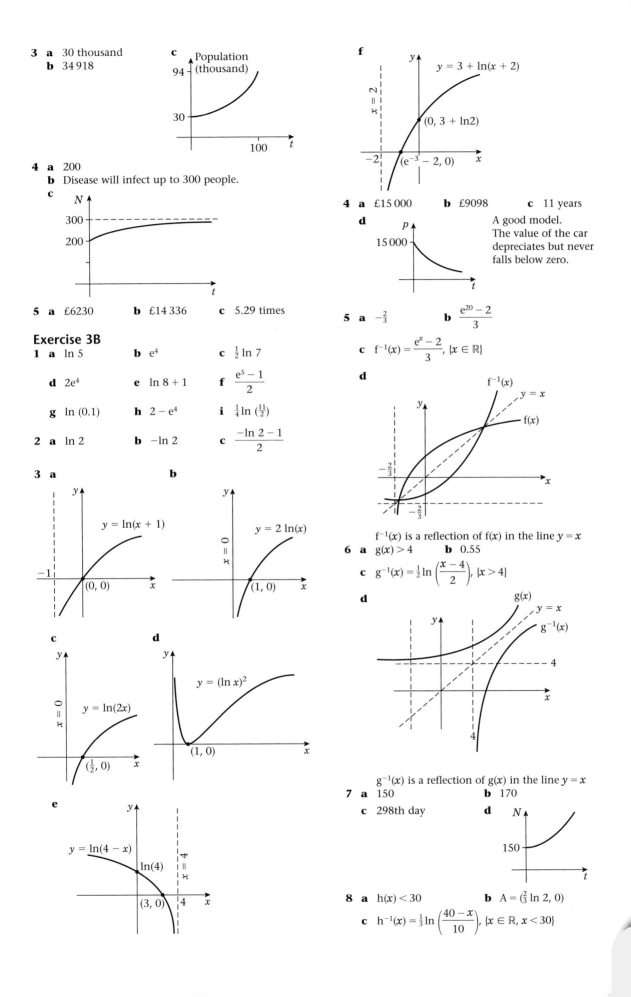

3 a 30 thousand
b 34 918
c Population (thousand)

4 a 200
b Disease will infect up to 300 people.
c

5 a £6230   **b** £14 336   **c** 5.29 times

Exercise 3B
1 a ln 5   **b** e^4 **c** $\frac{1}{2}\ln 7$

d $2e^4$ **e** ln 8 + 1   **f** $\dfrac{e^5 - 1}{2}$

g ln (0.1)   **h** $2 - e^4$ **i** $\frac{1}{4}\ln\left(\frac{11}{2}\right)$

2 a ln 2   **b** $-\ln 2$ **c** $\dfrac{-\ln 2 - 1}{2}$

3 a $y = \ln(x + 1)$

b $y = 2\ln(x)$

c $y = \ln(2x)$

d $y = (\ln x)^2$

e $y = \ln(4 - x)$

f $y = 3 + \ln(x + 2)$, $(0, 3 + \ln2)$, $(e^{-3} - 2, 0)$

4 a £15 000   **b** £9098   **c** 11 years
d
A good model.
The value of the car
depreciates but never
falls below zero.

5 a $-\frac{2}{3}$ **b** $\dfrac{e^{20} - 2}{3}$

c $f^{-1}(x) = \dfrac{e^x - 2}{3}$, $\{x \in \mathbb{R}\}$

d

$f^{-1}(x)$ is a reflection of $f(x)$ in the line $y = x$
6 a $g(x) > 4$ **b** 0.55

c $g^{-1}(x) = \frac{1}{2}\ln\left(\dfrac{x - 4}{2}\right)$, $\{x > 4\}$

d

$g^{-1}(x)$ is a reflection of $g(x)$ in the line $y = x$
7 a 150   **b** 170
c 298th day   **d**

8 a $h(x) < 30$ **b** $A = \left(\frac{2}{3}\ln 2, 0\right)$
c $h^{-1}(x) = \frac{1}{3}\ln\left(\dfrac{40 - x}{10}\right)$, $\{x \in \mathbb{R}, x < 30\}$

Mixed exercise 3C

1 a

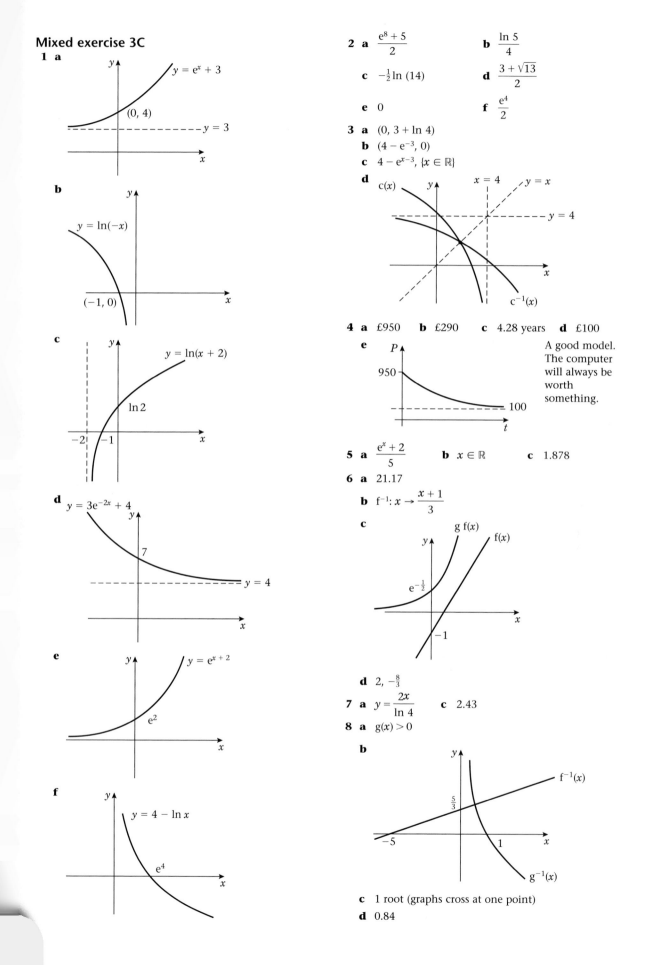

$y = e^x + 3$
$(0, 4)$
$y = 3$

b
$y = \ln(-x)$
$(-1, 0)$

c
$y = \ln(x + 2)$
$\ln 2$
-2 -1

d $y = 3e^{-2x} + 4$
7
$y = 4$

e $y = e^{x+2}$
e^2

f $y = 4 - \ln x$
e^4

2 a $\dfrac{e^8 + 5}{2}$ **b** $\dfrac{\ln 5}{4}$

c $-\tfrac{1}{2}\ln (14)$ **d** $\dfrac{3 + \sqrt{13}}{2}$

e 0 **f** $\dfrac{e^4}{2}$

3 a $(0, 3 + \ln 4)$
b $(4 - e^{-3}, 0)$
c $4 - e^{x-3}, \{x \in \mathbb{R}\}$
d

$c(x)$
$x = 4$
$y = x$
$y = 4$
$c^{-1}(x)$

4 a £950   **b** £290   **c** 4.28 years   **d** £100

e

P
950
100
t

A good model. The computer will always be worth something.

5 a $\dfrac{e^x + 2}{5}$ **b** $x \in \mathbb{R}$ **c** 1.878

6 a 21.17

b $f^{-1}: x \to \dfrac{x + 1}{3}$

c

$g\,f(x)$
$f(x)$
$e^{-\frac{1}{2}}$
-1

d $2, -\tfrac{8}{3}$

7 a $y = \dfrac{2x}{\ln 4}$ **c** 2.43

8 a $g(x) > 0$

b

$f^{-1}(x)$
$\tfrac{5}{3}$
-5
1
$g^{-1}(x)$

c 1 root (graphs cross at one point)
d 0.84

9 a $f(x) > k$

 b $2k$

 c $f^{-1}: x \to \ln(x - k), x > k$

 d
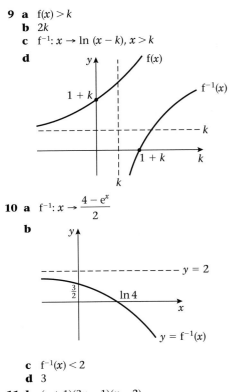

10 a $f^{-1}: x \to \dfrac{4 - e^x}{2}$

 b
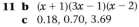

 c $f^{-1}(x) < 2$

 d 3

11 b $(x + 1)(3x - 1)(x - 2)$

 c 0.18, 0.70, 3.69

Exercise 4A

1 a $f(-2) = -1, f(-1) = 5$

 b $f(1) = 3, f(2) = -1$

 c $f(3) = -2.732, f(4) = 4$

 d $f(-0.5) = -0.125, f(-0.2) = 2.992$
   $f(1) = -2, f(2) = 5.5$

 e $f(-2) = -2, f(-1.8) = 0.264$
   $f(-1.8) = 0.264, f(-1) = -6$
   $f(2) = -18, f(3) = 98$

 f $f(2.2) = 0.020, f(2.3) = -0.087$

 g $f(1.65) = -0.294, f(1.75) = 0.195$

 h $f(0.5) = -0.084, f(0.6) = 0.018$

2 $f(5) = 2, f(4.9) = -0.401$

3 $f(1) = -1, f(2) = 1$

4 $f(0.5) = -0.210, f(0.6) = -0.029$

5 a

x	-3	-2	-1	0	1	2	3
$f(x)$	-1	11	11	5	-1	-1	11

 b -3

7 a
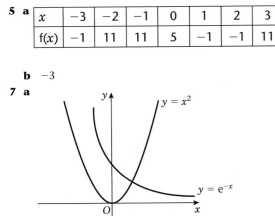

 b One point of intersection, so one root.

8 a
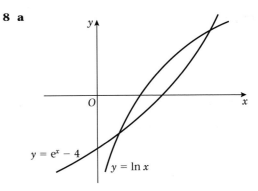

 b 2

9 a
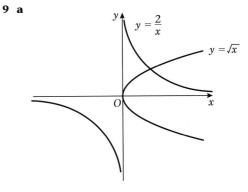

 b 1

 d $p = 3, q = 4$ **e** $4^{\frac{1}{3}}$

10 a
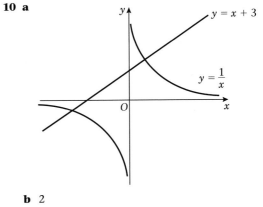

 b 2

 e 0.303

Exercise 4B

3 $\frac{4}{3}$

4 $\frac{1}{3}$

5 -1.5

6 1.104

7 1.20

8 a 6

 b 5.83

9 -1.72

10 1.653

Mixed exercise 4C

1 a 6, 2

b 2.646, 2.599, 2.602, 2.602

2 a

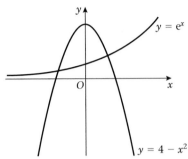

$y = e^x$

$y = 4 - x^2$

b -1.9646

c Iteration formula fails. Square root of a negative number.

3 $p = 5$, $q = 6$, $r = 5$; 1.708

4 a -0.065, 0.099

b Sign change

c $p = 0.8$, $q = 0.4$; 1.113, 1.118, 1.119, 1.120

5 a -0.0286, 0.279; Sign change

b $p = 1.5$, $q = -1$

c 0.4093

6 b 1.327

c $p = -1.25$

d -2.642

7 a 2.10

b $y^3 - 3y - 3 = 0$

d 0.68

8 a 1.5874, 1.5475, 1.5650, 1.5572, 1.5607

b Divergent sequence

9 a $q = -0.25$

b 0.1888

10 a

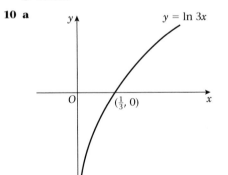

$y = \ln 3x$

$(\frac{1}{3}, 0)$

d 0.304

11 a

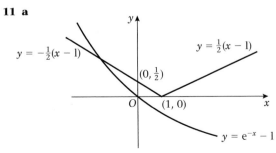

$y = -\frac{1}{2}(x - 1)$

$y = \frac{1}{2}(x - 1)$

$(0, \frac{1}{2})$

$(1, 0)$

$y = e^{-x} - 1$

d -0.6931, -0.6133

Review exercise 1

1 a $\dfrac{2x + 3}{x + 5}$ 　　**b** $x^2 - x + 1$

2 $\dfrac{4x - 3}{x(x - 3)}$

3 $\dfrac{x + 3}{x + 1}$

4 a $(2x + 1)(2x - 5)(4x - 1)$

b $(2x + 1)(2x - 5)$

5 a $\dfrac{x^2 + x + 1}{(x + 2)^2}$

b $(x + \frac{1}{2})^2 + \frac{3}{4} > 0$

c $x^2 + x + 1 > 0$ from **b** and $(x + 2)^2 > 0$ as $x \neq -2$

6 a $\dfrac{5 - 4x - x^2}{2(x + 1)^2}$ 　　**b** $x < -5$, $x > 1$

7 a $\dfrac{2(x - 4)}{2x - 1}$ 　　**b** $\dfrac{x - 8}{2(x - 1)}$

8 a

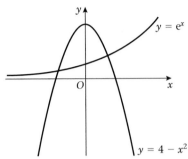

$y = f^{-1}(x)$

C' $(2, 5)$

$(0, 3)$

B'

$(-2, 0)$ A'

b i 2 　　**ii** $\sqrt{6}$

9 a $f^{-1}(x) = \dfrac{x - 4}{3}$

domain of $f^{-1}(x)$ is $x > 4$

b $\frac{11}{7}$

c range $g(x) > 1$

d $g^{-1}(x) = \dfrac{2x}{x - 1}$, domain $x > 1$

10 a $f(x) = \dfrac{2}{x - 1}$, $x > 1$

b $f^{-1}(x) = \dfrac{2 + x}{x}$ or $\left[1 + \dfrac{2}{x}\right]$

c $x = \pm 2$

11 a $f(x) \geqslant -16$

b For $f(x)$ to have an inverse function it must be one-to-one. With the given domain $f(x)$ is many-to-one.

c $\dfrac{64x}{(1 - x)^2}$

d $g^{-1}(x) = \dfrac{x - 8}{x}$

domain $x > 0$

12 a

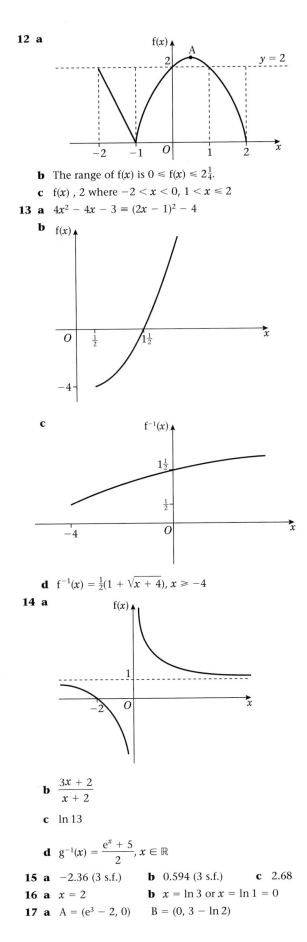

b The range of f(x) is $0 \leqslant f(x) \leqslant 2\frac{1}{4}$.

c f(x) , 2 where $-2 < x < 0$, $1 < x \leqslant 2$

13 a $4x^2 - 4x - 3 \equiv (2x - 1)^2 - 4$

b

c

d $f^{-1}(x) = \frac{1}{2}(1 + \sqrt{x + 4})$, $x \geqslant -4$

14 a

b $\dfrac{3x + 2}{x + 2}$

c ln 13

d $g^{-1}(x) = \dfrac{e^x + 5}{2}$, $x \in \mathbb{R}$

15 a -2.36 (3 s.f.)   **b** 0.594 (3 s.f.)   **c** 2.68

16 a $x = 2$ **b** $x = \ln 3$ or $x = \ln 1 = 0$

17 a $A = (e^3 - 2, 0)$ $B = (0, 3 - \ln 2)$

b

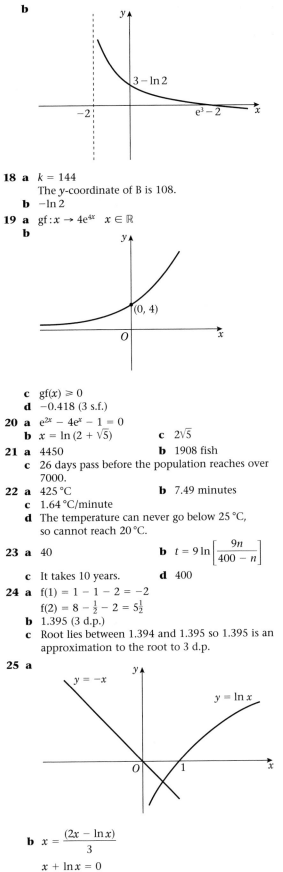

18 a $k = 144$

The y-coordinate of B is 108.

b $-\ln 2$

19 a $gf : x \to 4e^{4x}$ $x \in \mathbb{R}$

b

c $gf(x) \geqslant 0$

d -0.418 (3 s.f.)

20 a $e^{2x} - 4e^x - 1 = 0$

b $x = \ln(2 + \sqrt{5})$ **c** $2\sqrt{5}$

21 a 4450   **b** 1908 fish

c 26 days pass before the population reaches over 7000.

22 a 425 °C   **b** 7.49 minutes

c 1.64 °C/minute

d The temperature can never go below 25 °C, so cannot reach 20 °C.

23 a 40   **b** $t = 9\ln\left[\dfrac{9n}{400 - n}\right]$

c It takes 10 years.   **d** 400

24 a $f(1) = 1 - 1 - 2 = -2$

$f(2) = 8 - \frac{1}{2} - 2 = 5\frac{1}{2}$

b 1.395 (3 d.p.)

c Root lies between 1.394 and 1.395 so 1.395 is an approximation to the root to 3 d.p.

25 a

b $x = \dfrac{(2x - \ln x)}{3}$

$x + \ln x = 0$

c 0.56714 (5 d.p.)

26 a $f(1) = -0.6109\ldots$
$f(2) = 38.598\ldots$
b $x_1 = 1.131$ (3 d.p.)
$x_2 = 1.101$ (3 d.p.)
$x_3 = 1.088$ (3 d.p.)
c $x = 1.077$ (3 d.p.)
d 2.15 (2 d.p.)

27 a $x^5 - 5x - 4 = 0$ **b** $x = \sqrt[4]{\left(5 + \dfrac{4}{x}\right)}$

c $x_1 = 1.6640$ (5 s.f.)
$x_2 = 1.6495$
$x_3 = 1.6507$
$x_2 = 1.6506$
d $k = 1.6506$ (5 s.f.)

28 a $x = \sqrt{\left(\dfrac{2}{x} + \dfrac{1}{2}\right)}$

b $x_1 = 1.41$ (2 d.p.)
$x_2 = 1.39$ (2 d.p.)
$x_3 = 1.39$ (2 d.p.)
c $\alpha = 1.392$ (3 d.p.)

29 a $x \rightarrow \frac{1}{2}\ln\left[\dfrac{x+5}{4}\right]$

domain of f^{-1} is $x > -1$
b range of f^{-1} is $f^{-1}(x) > 0$
c $x_1 = 0.2814$
$x_2 = 0.2779$
d $x_3 = 0.2772$
$x_4 = 0.2771$
$x_5 = 0.2771$
e $k = 0.2771$ (4 d.p.)

30 a

b $x = \dfrac{1}{1 + x^2}$
c $x_1 = 0.6711$ (4 d.p.)
$x_2 = 0.6895$
$x_3 = 0.6778$
$x_4 = 0.6852$
d $k = 0.682$ (3 d.p.)

Exercise 5A

1 a

(1, 0), (0, 1)

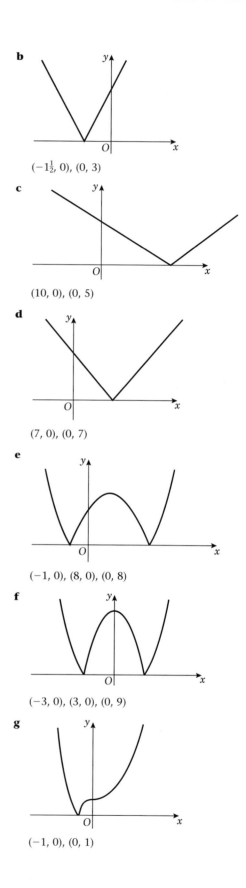

b

$(-1\frac{1}{2}, 0)$, (0, 3)

c

(10, 0), (0, 5)

d

(7, 0), (0, 7)

e

(−1, 0), (8, 0), (0, 8)

f

(−3, 0), (3, 0), (0, 9)

g

(−1, 0), (0, 1)

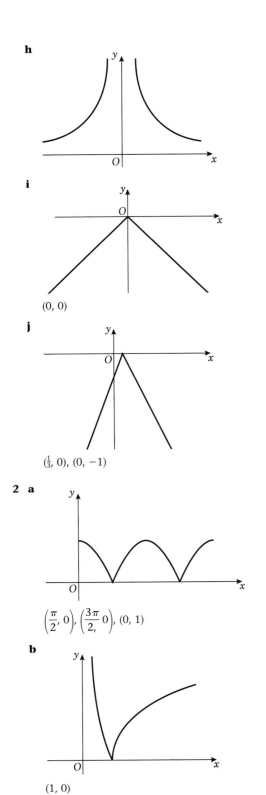

h

i

(0, 0)

j

$(\frac{1}{3}, 0)$, (0, −1)

2 a

$(\frac{\pi}{2}, 0)$, $(\frac{3\pi}{2}, 0)$, (0, 1)

b

(1, 0)

c

(1, 0), (0, 1)

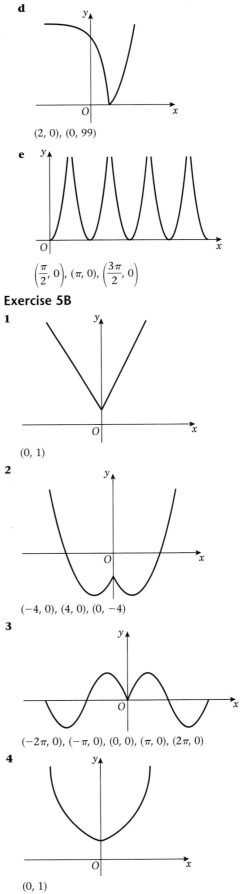

d

(2, 0), (0, 99)

e

$(\frac{\pi}{2}, 0)$, (π, 0), $(\frac{3\pi}{2}, 0)$

Exercise 5B

1

(0, 1)

2

(−4, 0), (4, 0), (0, −4)

3

(−2π, 0), (−π, 0), (0, 0), (π, 0), (2π, 0)

4

(0, 1)

Exercise 5C

1

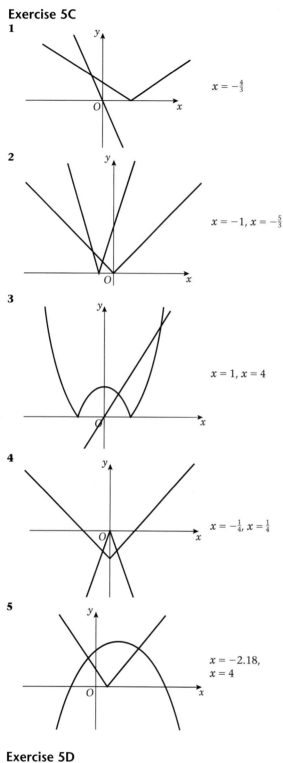

$x = -\frac{4}{3}$

2

$x = -1,\ x = -\frac{5}{3}$

3

$x = 1,\ x = 4$

4

$x = -\frac{1}{4},\ x = \frac{1}{4}$

5

$x = -2.18,$
$x = 4$

Exercise 5D

1 a

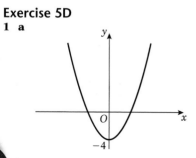

b

c

d

e

f

g

h

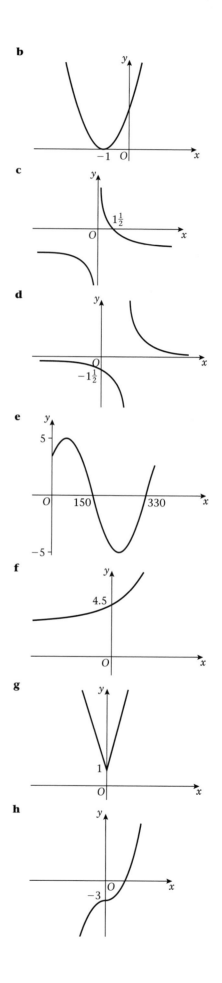

i

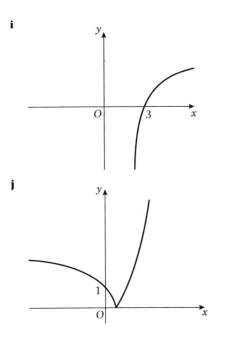

j

Exercise 5E

1 a

(0, 2)
(−2, −4)
(3, 14)

b

(2, −5)
(0, −7)
(5, −1)

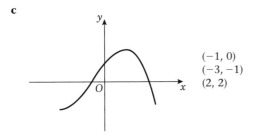

c

(−1, 0)
(−3, −1)
(2, 2)

d

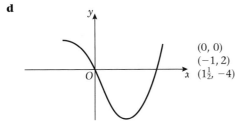

(0, 0)
(−1, 2)
(1½, −4)

2 a

(1, 12)
Max,
(0, 0)

b

(−2, 2)
Max,
(−4, 0)
(0, 1½)

c

(−1, 0)
Min,
(0, 1)

d

(−2, −8)
Min,
(−3, 0)

3 a

$x = 2$,
$y = -1$

b

$x = 0$,
$y = 4$

c

$x = 1$,
$y = 0$

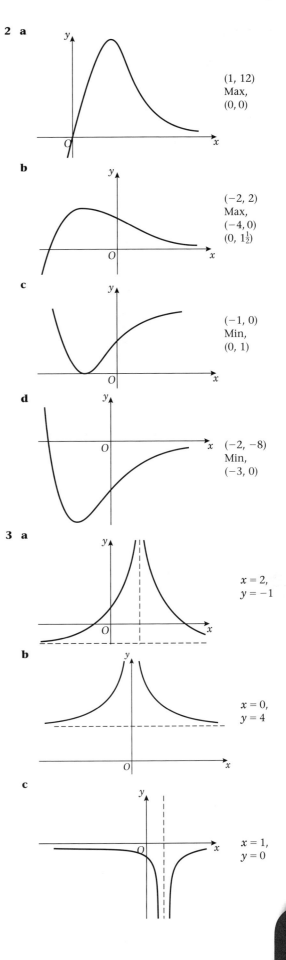

Mixed exercise 5F

1 a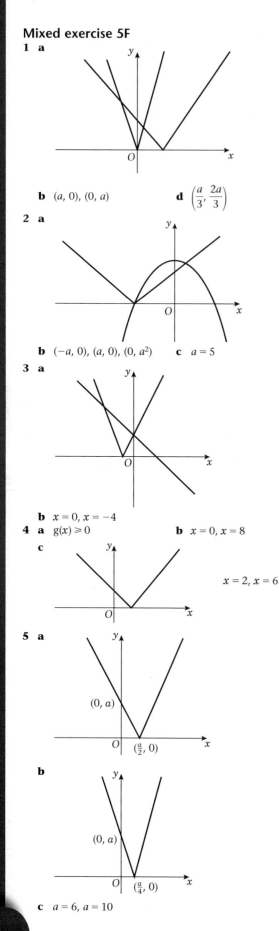

b $(a, 0), (0, a)$

d $\left(\dfrac{a}{3}, \dfrac{2a}{3}\right)$

2 a

b $(-a, 0), (a, 0), (0, a^2)$ **c** $a = 5$

3 a

b $x = 0, x = -4$

4 a $g(x) \geqslant 0$ **b** $x = 0, x = 8$

c

$x = 2, x = 6$

5 a

$(0, a)$

O $\left(\dfrac{a}{2}, 0\right)$

b

$(0, a)$

O $\left(\dfrac{a}{4}, 0\right)$

c $a = 6, a = 10$

6 a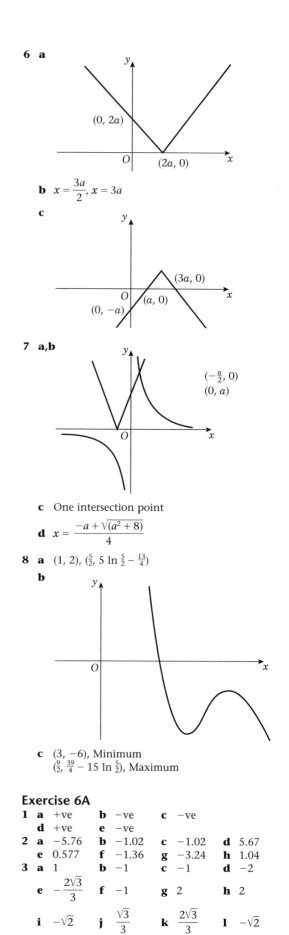

$(0, 2a)$

O $(2a, 0)$

b $x = \dfrac{3a}{2}, x = 3a$

c

$(3a, 0)$

O $(a, 0)$

$(0, -a)$

7 a,b

$\left(-\dfrac{a}{2}, 0\right)$

$(0, a)$

c One intersection point

d $x = \dfrac{-a + \sqrt{(a^2 + 8)}}{4}$

8 a $(1, 2), \left(\dfrac{5}{2}, 5\ln\dfrac{5}{2} - \dfrac{13}{4}\right)$

b

c $(3, -6)$, Minimum
$\left(\dfrac{9}{2}, \dfrac{39}{4} - 15\ln\dfrac{5}{2}\right)$, Maximum

Exercise 6A

1 a +ve   **b** −ve   **c** −ve
  d +ve   **e** −ve

2 a −5.76   **b** −1.02   **c** −1.02   **d** 5.67
  e 0.577   **f** −1.36   **g** −3.24   **h** 1.04

3 a 1   **b** −1   **c** −1   **d** −2
  e $-\dfrac{2\sqrt{3}}{3}$ **f** −1   **g** 2   **h** 2
  i $-\sqrt{2}$ **j** $\dfrac{\sqrt{3}}{3}$ **k** $\dfrac{2\sqrt{3}}{3}$ **l** $-\sqrt{2}$

4 a

θ	0°	30°	45°	60°	70°	80°	85°	95°
sec θ	1	1.15	1.41	2	2.92	5.76	11.47	−11.47

θ	100°	110°	120°	135°	150°	180°	210°
sec θ	−5.76	−2.92	−2	−1.41	−1.15	−1	−1.15

b

θ	10°	20°	30°	45°	60°	80°	90°	100°	120°
cosec θ	5.76	2.92	2	1.41	1.15	1.02	1	1.02	1.15

θ	135°	150°	160°	170°	190°	200°	210°	225°	240°
cosec θ	1.41	2	2.92	5.76	−5.76	−2.92	−2	−1.41	−1.15

θ	270°	300°	315°	330°	340°	350°	390°
cosec θ	−1	−1.15	−1.41	−2	−2.92	−5.76	2

c

θ	−90°	−60°	−45°	−30°	−10°	10°	30°	45°	60°
cot θ	0	−0.58	−1	−1.73	−5.67	5.67	1.73	1	0.58

θ	90°	120°	135°	150°	170°	210°	225°	240°	270°
cot θ	0	−0.58	−1	−1.73	−5.67	1.73	1	0.58	0

Exercise 6B

1 a i

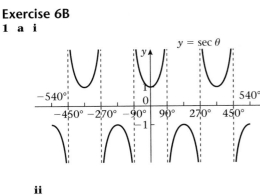

ii

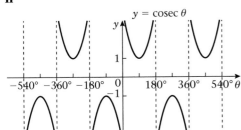

iii

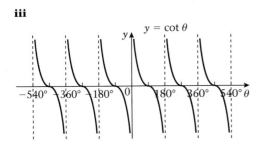

b i sec $\theta \leqslant -1$, sec $\theta \geqslant 1$
  ii cosec $\theta \leqslant -1$, cosec $\theta \geqslant 1$
  iii $\mathbb{R}$

2 a

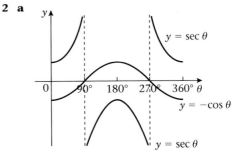

b The solutions of sec $\theta = -\cos \theta$ are the θ values of the points intersections of $y = \sec \theta$ and $y = -\cos \theta$. As they do not meet, there are no solutions.

3 a

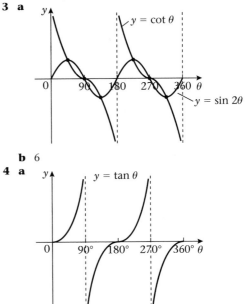

b 6

4 a

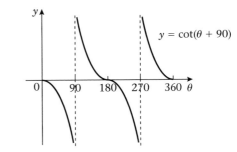

b $\cot(90 + \theta) = -\tan \theta$

5 a i The graph of $\tan\left(\theta + \dfrac{\pi}{2}\right)$ is the same as that of tan θ translated by $\dfrac{\pi}{2}$ to the left.

  ii The graph of $\cot(-\theta)$ is the same as that of cot θ reflected in the y-axis.

  iii The graph of $\operatorname{cosec}\left(\theta + \dfrac{\pi}{4}\right)$ is the same as that of cosec θ translated by $\dfrac{\pi}{4}$ to the left.

  iv The graph of $\sec\left(\theta - \dfrac{\pi}{4}\right)$ is the same as that of sec θ translated by $\dfrac{\pi}{4}$ to the right.

b $\tan\left(\theta + \dfrac{\pi}{2}\right) = \cot(-\theta)$; $\operatorname{cosec}\left(\theta + \dfrac{\pi}{4}\right) = \sec\left(\theta - \dfrac{\pi}{4}\right)$

6 a

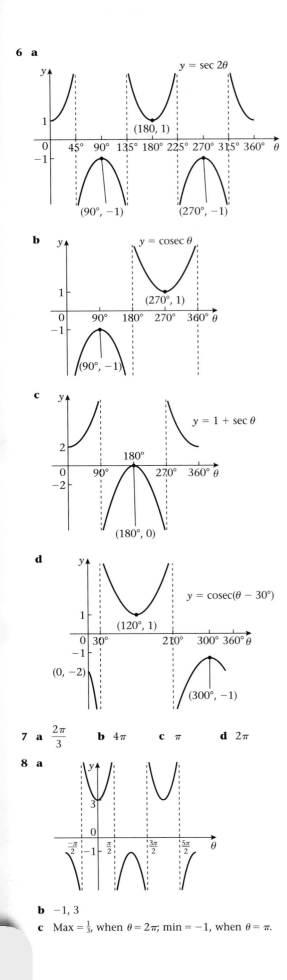

$y = \sec 2\theta$

(180, 1)

$45°$ $90°$ $135°$ $180°$ $225°$ $270°$ $315°$ $360°$ θ

(90°, −1) (270°, −1)

b $y = \mathrm{cosec}\,\theta$

(270°, 1)

(90°, −1)

c $y = 1 + \sec\theta$

180°

(180°, 0)

d $y = \mathrm{cosec}(\theta − 30°)$

(120°, 1)

(0, −2)

(300°, −1)

7 a $\dfrac{2\pi}{3}$ **b** 4π **c** π **d** 2π

8 a

b −1, 3

c Max $= \frac{1}{3}$, when $\theta = 2\pi$; min $= −1$, when $\theta = \pi$.

Exercise 6C

1 a $\mathrm{cosec}^3\theta$ **b** $2\cot^3\theta$ **c** $\frac{1}{2}\sec^2\theta$

   d $\cot^2\theta$ **e** $\sec^5\theta$ **f** $\mathrm{cosec}^2\theta$

   g $2\cot^{\frac{1}{2}}\theta$ **h** $\sec^3\theta$

2 a $\frac{5}{4}$ **b** $-\frac{1}{2}$ **c** $\pm\sqrt{3}$

3 a $\cos\theta$ **b** 1 **c** $\sec 2\theta$

   d 1 **e** 1 **f** $\cos A$

   g $\cos x$

4 a L.H.S. $= \cos\theta + \sin\theta\,\dfrac{\sin\theta}{\cos\theta} = \dfrac{\cos^2\theta + \sin^2\theta}{\cos\theta}$

$$= \dfrac{1}{\cos\theta} = \sec\theta = \text{R.H.S.}$$

b L.H.S. $= \dfrac{\cos\theta}{\sin\theta} + \dfrac{\sin\theta}{\cos\theta} \equiv \dfrac{\cos^2\theta + \sin^2\theta}{\sin\theta\cos\theta}$

$$\equiv \dfrac{1}{\sin\theta\cos\theta} = \dfrac{1}{\sin\theta} \times \dfrac{1}{\cos\theta}$$

$$\equiv \mathrm{cosec}\,\theta\sec\theta = \text{R.H.S.}$$

c L.H.S. $= \dfrac{1}{\sin\theta} - \sin\theta \equiv \dfrac{1 - \sin^2\theta}{\sin\theta} = \dfrac{\cos^2\theta}{\sin\theta}$

$$\equiv \cos\theta \times \dfrac{\cos\theta}{\sin\theta} \equiv \cos\theta\cot\theta = \text{R.H.S.}$$

d L.H.S. $= 1 - \cos x + \sec x - 1 \equiv \sec x - \cos x$

$$\equiv \dfrac{1}{\cos x} - \cos x \equiv \dfrac{1 - \cos^2 x}{\cos x} \equiv \dfrac{\sin^2 x}{\cos x}$$

$$\equiv \sin x \times \dfrac{\sin x}{\cos x} \equiv \sin x\tan x = \text{R.H.S.}$$

e L.H.S. $= \dfrac{\cos^2 x + (1 - \sin x)^2}{(1 - \sin x)\cos x}$

$$\equiv \dfrac{\cos^2 x + 1 - 2\sin x + \sin^2 x}{(1 - \sin x)\cos x}$$

$$\equiv \dfrac{2 - 2\sin x}{(1 - \sin x)\cos x} \equiv \dfrac{2(1 - \sin x)}{(1 - \sin x)\cos x}$$

$$\equiv 2\sec x = \text{R.H.S.}$$

f L.H.S. $\equiv \dfrac{\cos\theta}{1 + \dfrac{1}{\tan\theta}} \equiv \dfrac{\cos\theta}{\left(\dfrac{\tan\theta + 1}{\tan\theta}\right)}$

$$\equiv \dfrac{\cos\theta\tan\theta}{\tan\theta + 1} \equiv \dfrac{\sin\theta}{1 + \tan\theta} = \text{R.H.S.}$$

5 a $45°, 315°$ **b** $199°, 341°$

   c $112°, 292°$ **d** $30°, 150°$

   e $30°, 150°, 210°, 330°$ **f** $36.9°, 90°, 143°, 270°$

   g $26.6°, 207°$ **h** $45°, 135°, 225°, 315°$

6 a $90°$ **b** $\pm109°$

   c $-164°, 16.2°$ **d** $41.8°, 138°$

   e $\pm45°, \pm135°$ **f** $\pm60°$

   g $-173°, -97.2°, 7.24°, 82.8°$

   h $-152°, -36.5°, 28.4°, 143°$

7 a π **b** $\dfrac{5\pi}{6}, \dfrac{11\pi}{6}$ **c** $\dfrac{2\pi}{3}, \dfrac{4\pi}{3}$ **d** $\dfrac{\pi}{4}, \dfrac{3\pi}{4}$

8 a $\dfrac{AB}{AD} = \cos\theta \Rightarrow AD = 6\sec\theta$

$$\dfrac{AC}{AB} = \cos\theta \Rightarrow AC = 6\cos\theta$$

$$CD = AD - AC \Rightarrow CD = 6\sec\theta - 6\cos\theta$$

$$= 6(\sec\theta - \cos\theta)$$

b 2 cm

Exercise 6D

1 a $\sec^2(\frac{1}{2}\theta)$ **b** $\tan^2\theta$ **c** 1
d $\tan\theta$ **e** 1 **f** 3
g $\sin\theta$ **h** 1 **i** $\cos\theta$
j 1 **k** $4\csc^4 2\theta$

2 $\pm\sqrt{k-1}$

3 a $\frac{1}{2}$ **b** $-\dfrac{\sqrt{3}}{2}$

4 a $-\frac{5}{4}$ **b** $-\frac{4}{5}$ **c** $-\frac{3}{5}$

5 a $-\frac{7}{24}$ **b** $-\frac{25}{7}$

6 a L.H.S. $\equiv (\sec^2\theta - \tan^2\theta)(\sec^2\theta + \tan^2\theta)$
$\equiv 1(\sec^2\theta + \tan^2\theta) = $ R.H.S.
b L.H.S. $\equiv (1 + \cot^2 x) - (1 - \cos^2 x)$
$\equiv \cot^2 x + \cos^2 x = $ R.H.S.

c L.H.S. $\equiv \dfrac{1}{\cos^2 A}\left(\dfrac{\cos^2 A}{\sin^2 A} - \cos^2 A\right) \equiv \dfrac{1}{\sin^2 A} - 1$
$\equiv \csc^2 A - 1 = \cot^2 A = $ R.H.S.

d R.H.S. $\equiv \tan^2\theta \times \cos^2\theta \equiv \dfrac{\sin^2\theta}{\cos^2\theta} \times \cos^2\theta \equiv \sin^2\theta$
$\equiv 1 - \cos^2\theta = $ L.H.S.

e L.H.S. $= \dfrac{1 - \tan^2 A}{\sec^2 A} \equiv \cos^2 A\left(1 - \dfrac{\sin^2 A}{\cos^2 A}\right)$
$\equiv \cos^2 A - \sin^2 A$
$\equiv (1 - \sin^2 A) - \sin^2 A$
$\equiv 1 - 2\sin^2 A = $ R.H.S.

f L.H.S. $= \dfrac{1}{\cos^2\theta} + \dfrac{1}{\sin^2\theta} \equiv \dfrac{\sin^2\theta + \cos^2\theta}{\cos^2\theta\sin^2\theta}$
$\equiv \dfrac{1}{\cos^2\theta\sin^2\theta} \equiv \sec^2\theta\csc^2\theta = $ R.H.S.

g L.H.S. $= \csc A(1 + \tan^2 A) \equiv \csc A\left(1 + \dfrac{\sin^2 A}{\cos^2 A}\right)$
$\equiv \csc A + \dfrac{1}{\sin A}\cdot\dfrac{\sin^2 A}{\cos^2 A}$
$\equiv \csc A + \dfrac{\sin A}{\cos A}\cdot\dfrac{1}{\cos A}$
$\equiv \csc A + \tan A\sec A = $ R.H.S.

h L.H.S. $= \sec^2\theta - \sin^2\theta \equiv (1 + \tan^2\theta) - (1 - \cos^2\theta)$
$\equiv \tan^2\theta + \cos^2\theta \equiv $ R.H.S.

7 $\dfrac{\sqrt{2}}{4}$

8 a 20.9°, 69.1°, 201°, 249° **b** $\pm\dfrac{\pi}{3}$

c $-153°, -135°, 26.6°, 45°$ **d** $\dfrac{\pi}{2}, \dfrac{3\pi}{4}, \dfrac{3\pi}{2}, \dfrac{7\pi}{4}$

e 120° **f** $0, \dfrac{\pi}{4}, \pi$

g 0°, 180° **h** $\dfrac{\pi}{4}, \dfrac{\pi}{3}, \dfrac{5\pi}{4}, \dfrac{4\pi}{3}$

9 a $1 + \sqrt{2}$

b $\cos k° = \dfrac{1}{1 + \sqrt{2}} = \dfrac{\sqrt{2} - 1}{(\sqrt{2} - 1)(\sqrt{2} + 1)} = \sqrt{2} - 1$

c 65.5°, 294.5°

10 a $b = \dfrac{4}{a}$

b $c^2 = \cot^2 x = \dfrac{\cos^2 x}{\sin^2 x} = \dfrac{b^2}{1 - b^2} = \dfrac{\left(\frac{4}{a}\right)^2}{1 - \left(\frac{4}{a}\right)^2}$
$= \dfrac{16}{a^2} \times \dfrac{a^2}{(a^2 - 16)} = \dfrac{16}{a^2 - 16}$

11 a $\dfrac{1}{x} = \dfrac{1}{\sec\theta + \tan\theta} = \dfrac{\sec\theta - \tan\theta}{(\sec\theta - \tan\theta)(\sec\theta + \tan\theta)}$
$= \dfrac{\sec\theta - \tan\theta}{(\sec^2\theta - \tan^2\theta)} = \dfrac{\sec\theta - \tan\theta}{1}$

b $x^2 + \dfrac{1}{x^2} + 2 = \left(x + \dfrac{1}{x}\right)^2 = (2\sec\theta)^2 = 4\sec^2\theta$

12 $p = 2(1 + \tan^2\theta) - \tan^2\theta = 2 + \tan^2\theta$
$\Rightarrow \tan^2\theta = p - 2 \Rightarrow \cot^2\theta = \dfrac{1}{p - 2}$
$$\csc^2\theta = 1 + \cot^2\theta = 1 + \dfrac{1}{p - 2} = \dfrac{(p - 2) + 1}{p - 2} = \dfrac{p - 1}{p - 2}$$

Exercise 6E

1 a $\dfrac{\pi}{2}$ **b** $\dfrac{\pi}{2}$ **c** $-\dfrac{\pi}{4}$ **d** $-\dfrac{\pi}{6}$

e $\dfrac{3\pi}{4}$ **f** $-\dfrac{\pi}{6}$ **g** $\dfrac{\pi}{3}$ **h** $\dfrac{\pi}{3}$

2 a 0 **b** $-\dfrac{\pi}{3}$ **c** $\dfrac{\pi}{2}$

3 a $\frac{1}{2}$ **b** $-\frac{1}{2}$ **c** -1 **d** 0

4 a $\dfrac{\sqrt{3}}{2}$ **b** $\dfrac{\sqrt{3}}{2}$ **c** -1

d 2 **e** -1 **f** 1

5 $\alpha, \pi - \alpha$

6 a $0 < x < 1$

b i $\sqrt{1 - x^2}$ **ii** $\dfrac{x}{\sqrt{1 - x^2}}$

c i no change **ii** no change

7 a **b**

Range: $-\dfrac{\pi}{2} \leqslant f(x) \leqslant \dfrac{\pi}{2}$

c $g: x \to \arcsin 2x, -\frac{1}{2} \leqslant x \leqslant \frac{1}{2}$

d $g^{-1}: x \to \frac{1}{2}\sin x, -\dfrac{\pi}{2} \leqslant x \leqslant \dfrac{\pi}{2}$

8 a **b**

Range: $0 \leqslant \operatorname{arcsec} x \leqslant \pi, \operatorname{arcsec} x \neq \dfrac{\pi}{2}$

Mixed exercise 6F

1 $-125.3°, \pm54.7°$

2 $p = \dfrac{8}{q}$

3 $p = \sin\theta \Rightarrow \operatorname{cosec}\theta = \dfrac{1}{p}$; $q = 4\cot\theta \Rightarrow \cot\theta = \dfrac{q}{4}$

Using $1 + \cot^2\theta \equiv \operatorname{cosec}^2\theta$, $1 + \dfrac{q^2}{16} = \dfrac{1}{p^2}$

$\Rightarrow 16p^2 + p^2q^2 = 16 \Rightarrow p^2q^2 = 16 - 16p^2 = 16\,(1 - p^2)$

4 a i $60°$ **ii** $30°, 41.8°, 138.2°, 150°$

   b i $30°, 165°, 210°, 345°$ **ii** $45°, 116.6°, 225°, 296.6°$

   c i $\dfrac{71\pi}{60}, \dfrac{101\pi}{60}$ **ii** $\dfrac{\pi}{6}, \dfrac{5\pi}{6}, \dfrac{7\pi}{6}, \dfrac{11\pi}{6}$

5 $-\tfrac{8}{5}$

6 a L.H.S. $\equiv \left(\dfrac{\sin\theta}{\cos\theta} + \dfrac{\cos\theta}{\sin\theta}\right)(\sin\theta + \cos\theta)$

$\equiv \dfrac{(\sin^2\theta + \cos^2\theta)}{\cos\theta\sin\theta}(\sin\theta + \cos\theta)$

$\equiv \dfrac{\sin\theta}{\sin\theta\cos\theta} + \dfrac{\cos\theta}{\cos\theta\sin\theta}$

$\equiv \sec\theta + \operatorname{cosec}\theta \equiv$ R.H.S.

   b L.H.S. $\equiv \dfrac{\dfrac{1}{\sin x}}{\dfrac{1}{\sin x} - \sin x}$

$\equiv \dfrac{\dfrac{1}{\sin x}}{\dfrac{1 - \sin^2 x}{\sin x}}$

$\equiv \dfrac{1}{\sin x} \times \dfrac{\sin x}{\cos^2 x} \equiv \dfrac{1}{\cos^2 x} \equiv \sec^2 x \equiv$ R.H.S.

   c L.H.S. $\equiv 1 - \sin x + \operatorname{cosec} x - 1$

$\equiv -\sin x + \dfrac{1}{\sin x} \equiv \dfrac{1 - \sin^2 x}{\sin x} \equiv \dfrac{\cos^2 x}{\sin x}$

$\equiv \cos x\,\dfrac{\cos x}{\sin x} \equiv \cos x\cot x \equiv$ R.H.S.

   d L.H.S. $\equiv \dfrac{\cot x\,(1 + \sin x) - \cos x(\operatorname{cosec} x - 1)}{(\operatorname{cosec} x - 1)(1 + \sin x)}$

$\equiv \dfrac{\cot x + \cos x - \cot x + \cos x}{\operatorname{cosec} x - 1 + 1 - \sin x}$

$\equiv \dfrac{2\cos x}{\operatorname{cosec} x - \sin x} \equiv \dfrac{2\cos x}{\dfrac{1}{\sin x} - \sin x}$

$\equiv \dfrac{2\cos x}{\left(\dfrac{1 - \sin^2 x}{\sin x}\right)} \equiv \dfrac{2\cos x\sin x}{\cos^2 x}$

$\equiv 2\tan x \equiv$ R.H.S.

   e L.H.S. $\equiv \dfrac{\operatorname{cosec}\theta + 1 + \operatorname{cosec}\theta - 1}{(\operatorname{cosec}^2\theta - 1)} \equiv \dfrac{2\operatorname{cosec}\theta}{\cot^2\theta}$

$\equiv \dfrac{2}{\sin\theta} \cdot \dfrac{\sin^2\theta}{\cos^2\theta} \equiv \dfrac{2\sin\theta}{\cos^2\theta} \equiv \dfrac{2}{\cos\theta} \cdot \dfrac{\sin\theta}{\cos\theta}$

$\equiv 2\sec\theta\tan\theta \equiv$ R.H.S.

   f L.H.S. $\equiv \dfrac{\sec^2\theta - \tan^2\theta}{\sec^2\theta} \equiv \dfrac{1}{\sec^2\theta} \equiv \cos^2\theta \equiv$ R.H.S.

7 a L.H.S. $\equiv \dfrac{\sin^2 x + (1 + \cos x)^2}{(1 + \cos x)\sin x}$

$\equiv \dfrac{\sin^2 x + 1 + 2\cos x + \cos^2 x}{(1 + \cos x)\sin x}$

$\equiv \dfrac{2 + 2\cos x}{(1 + \cos x)\sin x} \equiv \dfrac{2(1 + \cos x)}{(1 + \cos x)\sin x}$

$\equiv \dfrac{2}{\sin x} \equiv 2\operatorname{cosec} x$

   b $-\dfrac{\pi}{3}, -\dfrac{2\pi}{3}, \dfrac{4\pi}{3}, \dfrac{5\pi}{3}$

8 R.H.S. $\equiv \left(\dfrac{1}{\sin\theta} + \dfrac{\cos\theta}{\sin\theta}\right)^2 \equiv \dfrac{(1 + \cos\theta)^2}{\sin^2\theta}$

$\equiv \dfrac{(1 + \cos\theta)^2}{1 - \cos^2\theta} \equiv \dfrac{(1 + \cos\theta)^2}{(1 - \cos\theta)(1 + \cos\theta)}$

$\equiv \dfrac{1 + \cos\theta}{1 - \cos\theta} \equiv$ L.H.S.

9 a $-2\sqrt{2}$

   b $\operatorname{cosec}^2 A = 1 + \cot^2 A = 1 + \tfrac{1}{8} = \tfrac{9}{8}$

$\Rightarrow \operatorname{cosec} A = \pm\dfrac{3}{2\sqrt{2}} = \pm\dfrac{3\sqrt{2}}{4}$

As A is obtuse, $\operatorname{cosec} A$ is +ve,

so $\Rightarrow \operatorname{cosec} A = \dfrac{3\sqrt{2}}{4}$

10 a $\dfrac{1}{k}$ **b** $k^2 - 1$

   c $-\dfrac{1}{\sqrt{k^2 - 1}}$ **d** $-\dfrac{k}{\sqrt{k^2 - 1}}$

11 $\dfrac{\pi}{12}, \dfrac{17\pi}{12}$

12 $\dfrac{\pi}{3}$

13 $\dfrac{\pi}{3}, \dfrac{5\pi}{6}, \dfrac{4\pi}{3}, \dfrac{11\pi}{6}$

14 a $(\sec x - 1)(\operatorname{cosec} x - 2)$

   b $0°, 30°, 150°, 360°$

15 $2 - \sqrt{3}$

16

17 a $-\tfrac{1}{3}$

   b i $-\tfrac{5}{3}$, **ii** $-\tfrac{4}{3}$ **c** $126.9°$

18 $pq = (\sec\theta - \tan\theta)(\sec\theta + \tan\theta) = \sec^2\theta - \tan^2\theta$

$= 1 \Rightarrow p = \dfrac{1}{q}$

19 a L.H.S. $\equiv (\sec^2\theta - \tan^2\theta)(\sec^2\theta + \tan^2\theta)$

$= 1 \times (\sec^2\theta + \tan^2\theta) = \sec^2\theta + \tan^2\theta$

$=$ R.H.S.

   b $-153.4°, -135°, 26.6°, 45°$

20 a

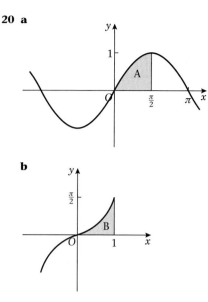

b

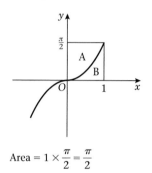

c The regions A and B fit together to make a rectangle.

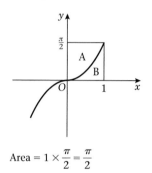

Area $= 1 \times \dfrac{\pi}{2} = \dfrac{\pi}{2}$

Exercise 7A

1 Example: With $A = 60°$, $B = 30°$,

$\sin(A + B) = \sin 90° = 1$; $\sin A + \sin B = \dfrac{\sqrt{3}}{2} + \dfrac{1}{2} \neq 1$

[You can find examples of A and B for which the statement is true, e.g. $A = 30°$, $B = -30°$, but one counter-example shows that it is not an identity.]

2 $\cos(\theta - \theta) \equiv \cos \theta \cos \theta + \sin \theta \sin \theta$
$\Rightarrow \sin^2 \theta + \cos^2 \theta \equiv 1$ as $\cos 0 = 0$

3 a $\sin\left(\dfrac{\pi}{2} - \theta\right) \equiv \sin \dfrac{\pi}{2} \cos \theta - \cos \dfrac{\pi}{2} \sin \theta$
$\equiv (1) \cos \theta - (0) \sin \theta = \cos \theta$

  b $\cos\left(\dfrac{\pi}{2} - \theta\right) \equiv \cos \dfrac{\pi}{2} \cos \theta + \sin \dfrac{\pi}{2} \sin \theta$
$\equiv (0) \cos \theta + (1) \sin \theta = \sin \theta$

4 a $\sin 35°$ **b** $\sin 35°$ **c** $\cos 210°$ **d** $\tan 31°$
  e $\cos \theta$ **f** $\cos 7\theta$ **g** $\sin 3\theta$ **h** $\tan 5\theta$
  i $\sin A$ **j** $\cos 3x$

5 a 1 **b** 0 **c** $\dfrac{\sqrt{3}}{2}$ **d** $\dfrac{\sqrt{2}}{2}$
  e $\dfrac{\sqrt{2}}{2}$ **f** $-\dfrac{1}{2}$ **g** $\sqrt{3}$ **h** $\dfrac{\sqrt{3}}{3}$
  i 1 **j** $\sqrt{2}$

6 a $(60 - \theta)°$

  b $\dfrac{\sin(60 - \theta)°}{3} = \dfrac{\sin \theta°}{4}$

$\Rightarrow 4(\sin 60° \cos \theta° - \cos 60° \sin \theta°) = 3 \sin \theta$

$\Rightarrow 4\left(\dfrac{\sqrt{3}}{2} \cos \theta° - \dfrac{1}{2} \sin \theta°\right) = 3 \sin \theta$

$\Rightarrow 2\sqrt{3} \cos \theta° = 5 \sin \theta° \Rightarrow \tan \theta° = \dfrac{2\sqrt{3}}{5}$

7 a L.H.S. $= \sin A \cos 60° + \cos A \sin 60°$
$\qquad\qquad + \sin A \cos 60° - \cos A \sin 60°$
$\qquad = 2 \sin A \cos 60°$
$\qquad = 2 \sin A \left(\dfrac{1}{2}\right) = \sin A = $ R.H.S.

  b L.H.S. $= \dfrac{\cos A \cos B - \sin A \sin B}{\sin B \cos B} = \dfrac{\cos (A + B)}{\sin B \cos B}$
$\qquad = $ R.H.S.

  c L.H.S. $= \dfrac{\sin x \cos y + \cos x \sin y}{\cos x \cos y}$
$\qquad = \dfrac{\sin x \cancel{\cos y}}{\cos x \cancel{\cos y}} + \dfrac{\cancel{\cos x} \sin y}{\cancel{\cos x} \cos y}$
$\qquad = \tan x + \tan y = $ R.H.S.

  d L.H.S. $= \dfrac{\cos x \cos y - \sin x \sin y}{\sin x \sin y} + 1$
$\qquad = \cot x \cot y - 1 + 1 = \cot x \cot y = $ R.H.S.

  e L.H.S. $= \cos \theta \cos \dfrac{\pi}{3} - \sin \theta \sin \dfrac{\pi}{3} + \sqrt{3} \sin \theta$
$\qquad = \dfrac{1}{2} \cos \theta - \dfrac{\sqrt{3}}{2} \sin \theta + \sqrt{3} \sin \theta$
$\qquad = \dfrac{1}{2} \cos \theta + \dfrac{\sqrt{3}}{2} \sin \theta$
$\qquad = \sin \dfrac{\pi}{6} \cos \theta + \cos \dfrac{\pi}{6} \sin \theta$
$\qquad = \sin \left(\dfrac{\pi}{6} + \theta\right) = $ R.H.S.

  f R.H.S. $= \dfrac{\dfrac{1}{\tan A} \dfrac{1}{\tan B} - 1}{\dfrac{1}{\tan A} + \dfrac{1}{\tan B}} = \dfrac{1 - \tan A \tan B}{\tan A + \tan B}$
$\qquad = \dfrac{1}{\dfrac{\tan A + \tan B}{1 - \tan A \tan B}} = \dfrac{1}{\tan (A + B)}$
$\qquad = \cot (A + B)$

  g L.H.S. $= [\sin 45° \cos \theta° + \cos 45° \sin \theta°]^2$
$\qquad + [\sin 45° \cos \theta° - \cos 45° \sin \theta°]^2$
$\qquad = \dfrac{1}{2} [(\cos \theta° + \sin \theta°)^2 + (\cos \theta° - \sin \theta°)^2]$
$\qquad = \dfrac{1}{2} [\cos^2 \theta° + \sin^2 \theta° + 2 \sin \theta° \cos \theta°$
$\qquad + \cos^2 \theta° + \sin^2 \theta° - 2 \sin \theta° \cos \theta°]$
$\qquad = \dfrac{1}{2} [1 + 1] = 1 = $ R.H.S.

  h L.H.S. $= [\cos A \cos B - \sin A \sin B] \times$
$\qquad [\cos A \cos B + \sin A \sin B]$
$\qquad = \cos^2 A \cos^2 B - \sin^2 A \sin^2 B$
$\qquad = \cos^2 A \cos^2 B - (1 - \cos^2 A) \sin^2 B$
$\qquad = \cos^2 A (\cos^2 B + \sin^2 B) - \sin^2 B$
$\qquad = \cos^2 A - \sin^2 B = $ R.H.S.

8 a $\dfrac{3 + 4\sqrt{3}}{10}$ **b** $\dfrac{4 + 3\sqrt{3}}{10}$ **c** $\dfrac{10(3\sqrt{3} - 4)}{11}$

9 a $\dfrac{3}{5}$ **b** $\dfrac{4}{5}$
  c $\dfrac{3 - 4\sqrt{3}}{10}$ **d** $\dfrac{1}{7}$

10 a $-\frac{77}{85}$ **b** $-\frac{36}{85}$ **c** $\frac{36}{77}$

11 a $-\frac{36}{325}$ **b** $\frac{204}{253}$ **c** $-\frac{325}{36}$

12 a $\cos 2\theta$ $[=\cos(\theta+\theta)]$ **b** $\sin 8\theta$ $[=\sin(4\theta+4\theta)]$

  c $\tan\left(\theta+\dfrac{\pi}{4}\right)$ **d** $\sin\left(\theta+\dfrac{\pi}{4}\right)$ or $\cos\left(\theta-\dfrac{\pi}{4}\right)$

13 a $51.7°, 231.7°$ **b** $170.1°, 350.1°$
  c $56.5°, 303.5°$ **d** $150°, 330°$
  e $153.4°, 161.6°, 333.4°, 341.6°$
  f $0°, 90°$

14 a $30°, 270°$ **b** $30°, 270°$

15 a $\tan(45+30)° = \dfrac{\tan 45° + \tan 30°}{1 - \tan 45° \tan 30°}$

  b $\tan 75 = \dfrac{1 + \dfrac{\sqrt{3}}{3}}{1 - \dfrac{\sqrt{3}}{3}} = \dfrac{3+\sqrt{3}}{3-\sqrt{3}} = \dfrac{(3+\sqrt{3})(3+\sqrt{3})}{(3-\sqrt{3})(3+\sqrt{3})}$

  $= \dfrac{12 + 6\sqrt{3}}{9 - 3} = 2 + \sqrt{3}$

16 $\cos 105° \quad = \cos(60+45)°$
  $= \cos 60° \cos 45° - \sin 60° \sin\ 45°$
  $= \left(\dfrac{1}{2}\right)\left(\dfrac{1}{\sqrt{2}}\right) - \left(\dfrac{\sqrt{3}}{2}\right)\left(\dfrac{1}{\sqrt{2}}\right) = \dfrac{1-\sqrt{3}}{2\sqrt{2}}$

  so $\sec 105° = \dfrac{2\sqrt{2}}{1-\sqrt{3}} = \dfrac{2\sqrt{2}(1+\sqrt{3})}{1-3} = -\sqrt{2}(1+\sqrt{3})$

17 a $\dfrac{\sqrt{2}(\sqrt{3}+1)}{4}$ **b** $\dfrac{\sqrt{2}(\sqrt{3}+1)}{4}$

  c $\dfrac{\sqrt{2}(\sqrt{3}-1)}{4}$ **d** $\sqrt{3}-2$

18 a $3(\sin x \cos y - \cos x \sin y)$
    $- (\sin x \cos y + \cos x \sin y) = 0$
    $\Rightarrow \quad 2\sin x \cos y - 4\cos x \sin y = 0$
    Divide throughout by $2\cos x \cos y$
    $\Rightarrow \tan x - 2\tan y = 0$, so $\tan x = 2\tan y$
  b Using **a** $\tan x = 2\tan y = 2\tan 45° = 2$
    so $x = 63.4°, 243.4°$

19 2

20 $\dfrac{\tan x - 3}{3\tan x + 1}$

21 a $\frac{5}{3}$ **b** $\sqrt{3}$ **c** $-\left(\dfrac{8+5\sqrt{3}}{11}\right)$

22 a 45   **b** 225

23 $\cos y = \sin x \cos y + \sin y \cos x$
  Divide by $\cos x \cos y \Rightarrow \sec x = \tan x + \tan y$,
  so $\tan y = \sec x - \tan x$

24 $-\frac{6}{7}$

25 $\dfrac{\tan x + \sqrt{3}}{1 - \sqrt{3}\tan x} = \dfrac{1}{2} \Rightarrow (2+\sqrt{3})\tan x = 1 - 2\sqrt{3}$, so

  $\tan x = \dfrac{1 - 2\sqrt{3}}{2 + \sqrt{3}} = \dfrac{(1 - 2\sqrt{3})(2 - \sqrt{3})}{1} = 8 - 5\sqrt{3}$

Exercise 7B

1 a $\sin 20°$ **b** $\cos 50°$ **c** $\cos 80°$
  d $\tan 10°$ **e** $\operatorname{cosec} 49°$ **f** $3\cos 60°$

  g $\frac{1}{2}\sin 16°$ **h** $\cos\left(\dfrac{\pi}{8}\right)$

2 a $\dfrac{\sqrt{2}}{2}$ **b** $\dfrac{\sqrt{3}}{2}$

  c $\frac{1}{2}$ **d** 1

3 a $\cos 6\theta$ **b** $3\sin 4\theta$ **c** $\tan\theta$
  d $2\cos\theta$ **e** $\sqrt{2}\cos\theta$ **f** $\frac{1}{4}\sin^2 2\theta$
  g $\sin 4\theta$ **h** $-\frac{1}{2}\tan 2\theta$ **i** $\cos^2 2\theta$

4 $-\frac{7}{8}$

5 $\pm\frac{1}{5}$

6 mn

7 a i $\frac{24}{7}$ **ii** $\frac{24}{25}$ **iii** $\frac{7}{25}$
  b $\frac{336}{625}$

8 a i $-\frac{7}{9}$ **ii** $\dfrac{2\sqrt{2}}{3}$ **iii** $-\dfrac{9\sqrt{2}}{8}$

  b $\tan 2A = \dfrac{\sin 2A}{\cos 2A} = -\dfrac{4\sqrt{2}}{9} \times -\dfrac{9}{7} = \dfrac{4\sqrt{2}}{7}$

9 -3

10 $51.3°$

11 a $\cos 2\theta = \dfrac{3^2 + 6^2 - 5^2}{2 \times 3 \times 6} = \dfrac{20}{36} = \dfrac{5}{9}$ **b** $\dfrac{\sqrt{2}}{3}$

12 a $\dfrac{3}{4}$

  b $m = \tan 2\theta = \dfrac{2\left(\dfrac{3}{4}\right)}{1 - \left(\dfrac{3}{4}\right)^2} = \dfrac{3}{2} \times \dfrac{16}{7} = \dfrac{24}{7}$

Exercise 7C

1 a L.H.S. $= \dfrac{\cos^2 A - \sin^2 A}{\cos A + \sin A}$

  $= \dfrac{(\cos A + \sin A)(\cos A - \sin A)}{\cos A + \sin A}$

  $= \cos A - \sin A = $ R.H.S.

  b R.H.S. $= \dfrac{2}{2\sin A \cos A}\{\sin B \cos A - \cos B \sin A\}$

  $= \dfrac{\sin B}{\sin A} - \dfrac{\cos B}{\cos A} = $ L.H.S.

  c L.H.S. $= \dfrac{1 - (1 - 2\sin^2\theta)}{2\sin\theta\cos\theta} = \dfrac{2\sin^2\theta}{2\sin\theta\cos\theta}$

  $= \tan\theta = $ R.H.S.

  d L.H.S. $= \dfrac{1 + \tan^2\theta}{1 - \tan^2\theta} = \dfrac{1 + \dfrac{\sin^2\theta}{\cos^2\theta}}{1 - \dfrac{\sin^2\theta}{\cos^2\theta}}$

  $= \dfrac{\cos^2\theta + \sin^2\theta}{\cos^2\theta - \sin^2\theta} = \dfrac{1}{\cos 2\theta} = \sec 2\theta = $ R.H.S.

  e L.H.S. $= 2\sin\theta\cos\theta\,(\sin^2\theta + \cos^2\theta)$
  $= 2\sin\theta\cos\theta = \sin 2\theta = $ R.H.S.

  f L.H.S. $= \dfrac{\sin 3\theta\cos\theta - \cos 3\theta\sin\theta}{\sin\theta\cos\theta}$

  $= \dfrac{\sin(3\theta - \theta)}{\sin\theta\cos\theta} = \dfrac{\sin 2\theta}{\sin\theta\cos\theta}$

  $= \dfrac{2\sin\theta\cos\theta}{\sin\theta\cos\theta} = 2 = $ R.H.S.

  g L.H.S. $= \dfrac{1}{\sin\theta} - \dfrac{2\cos 2\theta\cos\theta}{\sin 2\theta}$

  $= \dfrac{1}{\sin\theta} - \dfrac{2\cos 2\theta\cos\theta}{2\sin\theta\cos\theta} = \dfrac{1 - \cos 2\theta}{\sin\theta}$

  $= \dfrac{1 - (1 - 2\sin^2\theta)}{\sin\theta} = 2\sin\theta = $ R.H.S.

h L.H.S. $= \dfrac{\dfrac{1}{\cos\theta} - 1}{\dfrac{1}{\cos\theta} + 1} = \dfrac{1 - \cos\theta}{1 + \cos\theta}$

$= \dfrac{1 - \left(1 - 2\sin^2\dfrac{\theta}{2}\right)}{1 + \left(2\cos^2\dfrac{\theta}{2} - 1\right)} = \dfrac{2\sin^2\dfrac{\theta}{2}}{2\cos^2\dfrac{\theta}{2}} = \tan^2\dfrac{\theta}{2}$

$= $ R.H.S.

i L.H.S. $= \dfrac{1 - \tan x}{1 + \tan x} = \dfrac{\cos x - \sin x}{\cos x + \sin x}$

$= \dfrac{(\cos x - \sin x)(\cos x - \sin x)}{\cos^2 x - \sin^2 x}$

$= \dfrac{\cos^2 x + \sin^2 x - 2\sin x \cos x}{\cos^2 x - \sin^2 x}$

$= \dfrac{1 - \sin 2x}{\cos 2x} = $ R.H.S.

2 a L.H.S. $= \dfrac{\sin\theta}{\cos\theta} + \dfrac{\cos\theta}{\sin\theta} = \dfrac{\sin^2\theta + \cos^2\theta}{\sin\theta\cos\theta}$

$= \dfrac{1}{(\frac{1}{2})\sin 2\theta} = 2\operatorname{cosec} 2\theta = $ R.H.S.

b 4

3 a $0, \dfrac{\pi}{3}, \pi, \dfrac{5\pi}{3}, 2\pi$

b $\pm 38.7°$

c 30, 150, 210, 330

d $\dfrac{\pi}{12}, \dfrac{\pi}{4}, \dfrac{5\pi}{12}, \dfrac{3\pi}{4}$

e 33.2°, 146.8°, 213.2°, 326.8°

f 60°, 300°, 443.6°, 636.4°

g $\dfrac{\pi}{8}, \dfrac{5\pi}{8}$

h $\dfrac{\pi}{4}, \dfrac{5\pi}{4}$

i 0°, 30°, 150°, 180°, 210°, 330°

j $\dfrac{\pi}{6}, \dfrac{2\pi}{3}, \dfrac{7\pi}{6}, \dfrac{5\pi}{3}$

k 0°, ±113.6°, 180°

l −104.0°, 0°, 76.0°, 180°

m 0°, 35.3°, 144.7°, 180°, 215.3°, 324.7°

4 $q = \dfrac{p^2}{2} - 1$

5 a $y = 2(1 - x)$

b $2xy = 1 - x^2$

c $y^2 = 4x^2(1 - x^2)$

d $y^2 = \dfrac{2(4 - x)}{3}$

6 a L.H.S. $= \cos^2 2\theta + \sin^2 2\theta - 2\sin 2\theta\cos 2\theta$
$= 1 - \sin 4\theta = $ R.H.S.

b $\dfrac{\pi}{24}, \dfrac{17\pi}{24}$

7 a i R.H.S. $= \dfrac{2\tan\left(\dfrac{\theta}{2}\right)}{\sec^2\left(\dfrac{\theta}{2}\right)} = 2\dfrac{\sin\left(\dfrac{\theta}{2}\right)}{\cos\left(\dfrac{\theta}{2}\right)} \times \dfrac{\cos^2\left(\dfrac{\theta}{2}\right)}{1}$

$= 2\sin\left(\dfrac{\theta}{2}\right)\cos\left(\dfrac{\theta}{2}\right) = \sin\theta$

ii R.H.S. $= \dfrac{1 - \tan^2\left(\dfrac{\theta}{2}\right)}{1 + \tan^2\left(\dfrac{\theta}{2}\right)} = \dfrac{1 - \tan^2\left(\dfrac{\theta}{2}\right)}{\sec^2\left(\dfrac{\theta}{2}\right)}$

$= \cos^2\left(\dfrac{\theta}{2}\right)\left\{1 - \tan^2\left(\dfrac{\theta}{2}\right)\right\}$

$= \cos^2\left(\dfrac{\theta}{2}\right) - \sin^2\left(\dfrac{\theta}{2}\right) = \cos\theta = $ L.H.S.

b i 90°, 323.1°
ii 13.3°, 240.4°

8 a i $\cos x \equiv 2\cos^2\dfrac{x}{2} - 1$

$\Rightarrow 2\cos^2\dfrac{x}{2} \equiv 1 + \cos x \Rightarrow \cos^2\dfrac{x}{2} \equiv \dfrac{1 + \cos x}{2}$

ii $\cos x \equiv 1 - 2\sin^2\dfrac{x}{2}$

$\Rightarrow 2\sin^2\dfrac{x}{2} \equiv 1 - \cos x \Rightarrow \sin^2\dfrac{x}{2} \equiv \dfrac{1 - \cos x}{2}$

b i $\dfrac{2\sqrt{5}}{5}$ **ii** $\dfrac{\sqrt{5}}{5}$ **iii** $\dfrac{1}{2}$

c $\cos^4\dfrac{A}{2} \equiv \left(\dfrac{1 + \cos A}{2}\right)^2 \equiv \dfrac{1 + 2\cos A + \cos^2 A}{4}$

$\equiv \dfrac{1 + 2\cos A + \left(\dfrac{1 + \cos 2A}{2}\right)}{4}$

$\equiv \dfrac{2 + 4\cos A + 1 + \cos 2A}{8}$

$\equiv \dfrac{3 + 4\cos A + \cos 2A}{8}$

9 a L.H.S. $\equiv \dfrac{3(1 + \cos 2x)}{2} - \dfrac{(1 - \cos 2x)}{2}$

$\equiv 1 + 2\cos 2x$

b

Crosses y-axis at $(0, 3)$

Crosses x-axis at

$\left(-\dfrac{2\pi}{3}, 0\right), \left(-\dfrac{\pi}{3}, 0\right), \left(\dfrac{\pi}{3}, 0\right), \left(\dfrac{2\pi}{3}, 0\right)$

10 a $2\cos^2\left(\dfrac{\theta}{2}\right) - 4\sin^2\left(\dfrac{\theta}{2}\right) = 2\left(\dfrac{1+\cos\theta}{2}\right) - 4\left(\dfrac{1-\cos\theta}{2}\right)$

$$= 1 + \cos\theta - 2 + 2\cos\theta$$
$$= 3\cos\theta - 1$$

b $131.8°, 228.2°$

11 a $(\sin^2 A + \cos^2 A)^2 \equiv \sin^4 A + \cos^4 A + 2\sin^2 A \cos^2 A$

So $\quad 1 \equiv \sin^4 A + \cos^4 A + \dfrac{(2\sin A \cos A)^2}{2}$

$\Rightarrow \quad 2 \equiv 2(\sin^4 A + \cos^4 A) + \sin^2 2A$

$\sin^4 A + \cos^4 A \equiv \tfrac{1}{2}(2 - \sin^2 2A)$

b Using **a**: $\sin^4 A + \cos^4 A \equiv \tfrac{1}{2}(2 - \sin^2 2A)$

$$\equiv \tfrac{1}{2}\left\{ 2 - \dfrac{(1 - \cos 4A)}{2} \right\}$$
$$\equiv \dfrac{(4 - 1 + \cos 4A)}{4}$$
$$\equiv \dfrac{3 + \cos 4A}{4}$$

c $\dfrac{\pi}{12}, \dfrac{5\pi}{12}, \dfrac{7\pi}{12}, \dfrac{11\pi}{12}$

12 a $\cos(2A + A) \equiv \cos 2A \cos A - \sin 2A \sin A$

$$\equiv (\cos^2 A - \sin^2 A)\cos A - 2\sin^2 A \cos A$$
$$\equiv \cos^3 A - 3\sin^2 A \cos A$$
$$\equiv \cos^3 A - 3(1 - \cos^2 A)\cos A$$
$$\equiv \cos^3 A - 3\cos A + 3\cos^3 A$$
$$\equiv 4\cos^3 A - 3\cos A$$

b $20°, 100°, 140°, 220°, 260°, 340°$

13 a $\tan 3\theta \equiv \dfrac{\sin 3\theta}{\cos 3\theta} = \dfrac{3\sin\theta\cos^2\theta - \sin^3\theta}{\cos^3\theta - 3\sin^2\theta\cos\theta}$

> Use the intermediate step result, that includes $\sin\theta$ and $\cos\theta$, for $\sin 3\theta$ and $\cos 3\theta$. See Example 13 for $\sin 3\theta$, and the solution to question 12 for $\cos 3\theta$.

$$= \dfrac{3\tan\theta - \tan^3\theta}{1 - 3\tan^2\theta}$$

> Divide 'top and bottom' by $\cos^3\theta$.

b

$\tan\theta = 2\sqrt{2}$

so $\tan 3\theta = \dfrac{6\sqrt{2} - 16\sqrt{2}}{1 - 24} = \dfrac{-10\sqrt{2}}{-23} = \dfrac{10\sqrt{2}}{23}$

Exercise 7D

1 $R = 13;\ \tan\alpha = \dfrac{12}{5}$

2 $35.3°$

3 $41.8°$

---

4 a R.H.S. $= \sqrt{2}\left\{\sin\theta\cos\dfrac{\pi}{4} + \cos\theta\sin\dfrac{\pi}{4}\right\}$

$$= \sqrt{2}\left\{\sin\theta\ \dfrac{1}{\sqrt{2}} + \cos\theta\ \dfrac{1}{\sqrt{2}}\right\}$$
$$= \sin\theta + \cos\theta = \text{L.H.S.}$$

b R.H.S. $= 2\left\{\sin 2\theta\cos\dfrac{\pi}{6} - \cos 2\theta\sin\dfrac{\pi}{6}\right\}$

$$= 2\left\{\sin 2\theta\ \dfrac{\sqrt{3}}{2} - \cos 2\theta\ \dfrac{1}{2}\right\}$$
$$= \sqrt{3}\sin 2\theta - \cos 2\theta = \text{L.H.S.}$$

5 $2\cos\left(2\theta + \dfrac{\pi}{3}\right) \equiv 2\left(\cos 2\theta\cos\dfrac{\pi}{3} - \sin 2\theta\sin\dfrac{\pi}{3}\right)$

$$\equiv 2\left(\cos 2\theta\ \dfrac{1}{2} - \sin 2\theta\ \dfrac{\sqrt{3}}{2}\right)$$
$$\equiv \cos 2\theta - \sqrt{3}\sin 2\theta$$

The other part follows from **4b**

6 a $R = \sqrt{10},\ \alpha = 71.6°$ **b** $R = 5,\ \alpha = 53.1°$
   c $R = \sqrt{53},\ \alpha = 74.1°$ **d** $R = \sqrt{5},\ \alpha = 63.4°$

7 a $\cos\theta - \sqrt{3}\sin\theta \equiv R\cos(\theta + \alpha)$ gives $R = 2,\ \alpha = \dfrac{\pi}{3}$

b $y = 2\cos\left(\theta + \dfrac{\pi}{3}\right)$

> The graph of $y = \cos\theta$ translated by $\dfrac{\pi}{3}$ to the left and stretched by factor of 2 vertically.

8 a $3\sin 3\theta - 4\cos 3\theta \equiv R\sin(3\theta - \alpha)$
$$\equiv R\sin 3\theta\cos\alpha - R\cos 3\theta\sin\alpha$$
So $R\cos\alpha \equiv 3,\ R\sin\alpha \equiv 4 \Rightarrow R = 5,$

$\tan\alpha = \dfrac{4}{3}$ so $\alpha = \tan^{-1}\dfrac{4}{3} = 53.1°$

b Minimum value is -5,
when $3\theta - 53.1 = 270 \Rightarrow \theta = 107.7$

9 a $\sqrt{2}\sin\left(2\theta + \dfrac{\pi}{4}\right)$ **b** $0, \dfrac{\pi}{4}, \pi, \dfrac{5\pi}{4}$

10 a $25\cos(\theta + 73.7°)$ **b** $(0, 7)$
   c $25, -25$ **d i** 2 **ii** 0 **iii** 1

11 a $5\left(\dfrac{1 - \cos 2\theta}{2}\right) - 3\left(\dfrac{1 + \cos 2\theta}{2}\right) + 3\sin 2\theta$

$$\equiv 1 + 3\sin 2\theta - 4\cos 2\theta, \text{ so } a = 3,\ b = -4,\ c = 1$$

b Maximum $= 6$, minimum $= -4$

12 a $6.9°, 66.9°$ **b** $16.6°, 65.9°$
   c $8.0°, 115.9°$ **d** $-165.2°, 74.8°$

13 a $39.1°, 152.2°, 219.1°, 332.2°$
   b $126.9°$ **c** $31.7°, 121.7°$ **d** $\dfrac{\pi}{3}$

14 a $125.3°, 305.3°$ **b** $109.5°$ **c** $70.5°$
No solutions for **d** and **e**

15 a $R = \sqrt{10},\ \alpha = 18.4°,\ \theta = 69.2°, 327.7°$
   b $9\cos^2\theta = 4 - 4\sin\theta + \sin^2\theta$
$$\Rightarrow 9(1 - \sin^2\theta) = 4 - 4\sin\theta + \sin^2\theta$$
So $10\sin^2\theta - 4\sin\theta - 5 = 0$
   c $69.2°, 110.8°, 212.3°, 327.7°$
   d When you square you are also solving
$3\cos\theta = -(2 - \sin\theta)$. The other two solutions are for this equation.

Exercise 7E

1 c i $\sin 9\theta + \sin 5\theta$ **ii** $\sin 17\theta + \sin 7\theta$

  d $0°, 90°, 180°$

  e $\dfrac{2\sin 4\theta \cos 3\theta}{2\sin 4\theta \cos \theta} \equiv \dfrac{\cos 3\theta}{\cos \theta}$

2 b i $\sin 8x - \sin 2x$

   ii $\frac{1}{2}(\sin 3x - \sin x)$

   iii $3(\sin 2x - \sin x)$

  d $\sin 56° - \sin 34° = 2\cos 45° \sin 11° = \sqrt{2}\sin 11°$

3 b i $\cos 3\theta + \cos 2\theta$

   ii $\frac{5}{2}(\cos 5x + \cos x)$

  d $\dfrac{2\cos 2\theta \sin \theta}{2\cos 2\theta \cos \theta} = \tan \theta$

4 d $0°, 30°, 90°, 150°, 180°$

5 a $\sin 10x + \sin 6x$

  b $\frac{1}{2}(\cos 6x + \cos 4x)$

  c $-\frac{3}{2}(\cos 8x - \cos 6x)$

  d $\frac{1}{2}(\cos 140° + \cos 60°)$

  e $5(\sin 2x - \sin x)$

  f $\sin 40° + \sin 20°$

6 $\sin 120° + \sin 45° = \dfrac{\sqrt{3}}{2} + \dfrac{\sqrt{2}}{2} = \dfrac{1}{2}(\sqrt{3} + \sqrt{2})$

7 a $2\sin 10x \cos 2x$

  b $-2\sin 2y \sin x$

  c $\cos 3x \sin 2x$

  d $\sqrt{2}\cos 50°$

  e $2\cos\left(\dfrac{9\pi}{120}\right)\cos\left(\dfrac{-\pi}{120}\right) = 2\cos\left(\dfrac{9\pi}{120}\right)\cos\left(\dfrac{\pi}{120}\right)$

  f $2\sin 85° \cos 65°$

8 $\cos\theta + \cos\left(\theta + \dfrac{4\pi}{3}\right) + \cos\left(\theta + \dfrac{2\pi}{3}\right)$

$= 2\cos\left(\theta + \dfrac{2\pi}{3}\right)\cos\left(\dfrac{2\pi}{3}\right) + \cos\left(\theta + \dfrac{2\pi}{3}\right)$

$= \cos\left(\theta + \dfrac{2\pi}{3}\right)\left\{2\cos\left(\dfrac{2\pi}{3}\right) + 1\right\}$

$= \cos\left(\theta + \dfrac{2\pi}{3}\right)\{0\} = 0$

9 $\dfrac{2\sin 45° \cos 30°}{-2\sin 45° \sin(-30°)} = \dfrac{1}{\tan 30°} = \sqrt{3}$

10 a $0°, 60°, 120°, 180°$ **b** $0, \dfrac{\pi}{4}, \dfrac{3\pi}{4}, \pi, \dfrac{5\pi}{4}, \dfrac{7\pi}{4}, 2\pi$

  c $25°, 145°$ **d** $\dfrac{\pi}{6}, \dfrac{\pi}{4}, \dfrac{3\pi}{4}, \dfrac{5\pi}{6}, \dfrac{5\pi}{4}, \dfrac{7\pi}{4}$

11 a $\dfrac{\sin 7\theta - \sin 3\theta}{\sin \theta \cos \theta} \equiv \dfrac{2\cos 5\theta \sin 2\theta}{\frac{1}{2}\sin 2\theta} \equiv 4\cos 5\theta$

  b $\dfrac{\cos 2\theta + \cos 4\theta}{\sin 2\theta - \sin 4\theta} \equiv \dfrac{2\cos 3\theta \cos \theta}{2\cos 3\theta \sin(-\theta)} \equiv -\cot \theta$

  c $[\sin(x+y) + \sin(x-y)][\sin(x+y) - \sin(x-y)]$
$\equiv [2\sin x \cos y][2\cos x \sin y]$
$\equiv [2\sin x \cos x][2\cos y \sin y]$
$\equiv \sin 2x \sin 2y$

  d $\cos 5x + \cos x + 2\cos 3x$
$\equiv 2\cos 3x \cos 2x + 2\cos 3x$
$\equiv 2\cos 3x (\cos 2x + 1)$
$\equiv 2\cos 3x (2\cos^2 x)$
$\equiv 4\cos^2 x \cos 3x$

12 a $\cos\theta - \cos 3\theta + \sin 2\theta$
$\equiv -2\sin 2\theta \sin(-\theta) + \sin 2\theta$
$\equiv 2\sin 2\theta \sin \theta + \sin 2\theta \equiv \sin 2\theta(2\sin\theta + 1)$

  b $0, \dfrac{\pi}{2}, \pi, \dfrac{7\pi}{6}, \dfrac{3\pi}{2}, \dfrac{11\pi}{6}, 2\pi$

Mixed exercise 7F

1 a $\tan A = 2, \tan B = \frac{1}{3}$

  b $45°$

2 $\sin x = \dfrac{1}{\sqrt{5}}$, so $\cos x = \dfrac{2}{\sqrt{5}}$

$\cos(x-y) = \sin y \Rightarrow \dfrac{2}{\sqrt{5}}\cos y + \dfrac{1}{\sqrt{5}}\sin y = \sin y$

$\Rightarrow (\sqrt{5} - 1)\sin y = 2\cos y$

$\Rightarrow \tan y = \dfrac{2}{\sqrt{5} - 1} = \dfrac{2(\sqrt{5} + 1)}{4}$

$= \dfrac{\sqrt{5} + 1}{2}$

3 a Seting $\theta = \dfrac{\pi}{8}$ gives resulting quadratic equation in t, $t^2 + 2t - 1 = 0$, where $t = \tan\left(\dfrac{\pi}{8}\right)$.

Solving this and taking +ve value for t gives result.

  b Expanding $\tan\left(\dfrac{\pi}{4} + \dfrac{\pi}{8}\right)$ gives answer: $\sqrt{2} + 1$

4 Sine rule: $\dfrac{\sin(\theta + 30°)}{5} = \dfrac{\sin(\theta - 30°)}{4}$.

Expand and use exact forms of $\sin 30°$ and $\cos 30°$.

5 a i $\frac{56}{65}$ **ii** $\frac{120}{119}$

  b Use $\cos\{180° - (A + B)\} \equiv -\cos(A + B)$ and expand. You can work out all the required trig. ratios (A *and* B are acute).

6 a L.H.S. $= \dfrac{1}{\cos\theta} \cdot \dfrac{1}{\sin\theta} \equiv \dfrac{1}{\frac{1}{2}\sin 2\theta}$
$\equiv 2\operatorname{cosec} 2\theta = $ R.H.S.

  b L.H.S. $= \dfrac{1 - (1 - 2\sin^2 x)}{1 + (2\cos^2 x - 1)} \equiv \dfrac{2\sin^2 x}{2\cos^2 x}$
$\equiv \tan^2 x = \sec^2 x - 1 = $ R.H.S.

  c L.H.S. $= \dfrac{\cos\theta}{\sin\theta} - \dfrac{2\cos 2\theta}{\sin 2\theta}$
$\equiv \dfrac{\cos\theta}{\sin\theta} - \dfrac{2\cos 2\theta}{2\sin\theta\cos\theta} \equiv \dfrac{\cos^2\theta - \cos 2\theta}{\sin\theta\cos\theta}$
$\equiv \dfrac{\cos^2\theta - (\cos^2\theta - \sin^2\theta)}{\sin\theta\cos\theta} \equiv \dfrac{\sin^2\theta}{\sin\theta\cos\theta}$
$\equiv \dfrac{\sin\theta}{\cos\theta} \equiv \tan\theta = $ R.H.S.

  d L.H.S. $= \cos^4 2\theta - \sin^4 2\theta$
$\equiv (\cos^2 2\theta - \sin^2 2\theta)(\cos^2 2\theta + \sin^2 2\theta)$
$\equiv (\cos^2 2\theta - \sin^2 2\theta)(1)$
$\equiv \cos 4\theta = $ R.H.S.

  e L.H.S. $= \dfrac{1 + \tan x}{1 - \tan x} - \dfrac{1 - \tan x}{1 + \tan x}$

$\equiv \dfrac{(1 + \tan x)^2 - (1 - \tan x)^2}{(1 + \tan x)(1 - \tan x)}$

$\equiv \dfrac{(1 + 2\tan x + \tan^2 x) - (1 - 2\tan x + \tan^2 x)}{1 - \tan^2 x}$

$\equiv \dfrac{4\tan x}{1 - \tan^2 x} = \dfrac{2(2\tan x)}{1 - \tan^2 x} = 2\tan 2x = $ R.H.S.

  f L.H.S. $= -\frac{1}{2}[\cos 2x - \cos 2y]$
$\equiv \frac{1}{2}[\cos 2y - \cos 2x]$
$\equiv \frac{1}{2}[2\cos^2 y - 1 - (2\cos^2 x - 1)]$
$\equiv \frac{1}{2}[2\cos^2 y - 2\cos^2 x]$
$\equiv \cos^2 y - \cos^2 x = $ R.H.S.

g L.H.S. $= 2\cos 2\theta + 1 + (2\cos^2 2\theta - 1)$
$\qquad\quad \equiv 2\cos 2\theta(1 + \cos 2\theta)$
$\qquad\quad \equiv 2\cos 2\theta(2\cos^2 \theta)$
$\qquad\quad \equiv 4\cos^2 \theta \cos 2\theta$
$\qquad\quad = $ R.H.S.

7 a Use $\cos 2x \equiv 1 - 2\sin^2 x$ **b** $\frac{4}{5}$

c i Use $\tan x = 2$, $\tan y = \frac{1}{3}$ in the expansion of $\tan(x + y)$.

  ii Find $\tan(x - y) = 1$ and note that $x - y$ has to be acute.

8 a Show that both sides are equal to $\frac{5}{6}$.

b $\dfrac{3k}{2}$ **c** $\dfrac{12k}{4 - 9k^2}$

9 a $\dfrac{\pi}{12}, \dfrac{7\pi}{12}$ **b** $0, \pi, 2\pi$

c $135°, 315°$ **d** $0°, 151.9°, 360°$

e $72.4°, 220.2°$ **f** $0, \dfrac{\pi}{4}, \dfrac{\pi}{2}, \dfrac{3\pi}{4}, \pi$

g $0.187, 2.95$

10 $75°$

11 $19.5°, 130.5°, 199.5°, 310.5°$

12 a Max $= 1$, $\theta = 100°$; Min $= -1$, $\theta = 280°$
  b Max $= 1$, $\theta = 330°$; Min $= -1$, $\theta = 150°$
  c Max $= \sqrt{2}$, $\theta = 45°$; Min $= -\sqrt{2}$, $\theta = 225°$

13 a $2\sin(x - 60)°$

b

Graph crossses y-axis at $(0, -\sqrt{3})$
Graph crossses x-axis at $(-300°, -0), (-120°, 0),$
$(60°, 0), (240°, 0)$

14 a $R = 25$, $\alpha = 1.29$ **b** 32

15 a Find $\sin \alpha = \frac{3}{5}$ and $\cos \alpha = \frac{4}{5}$ and insert in expansions on L.H.S. Result follows.
  b $0.6, 0.8$

16 a i $\frac{1}{2}$ **ii** $\frac{1}{2}$ **iii** $\dfrac{\sqrt{3}}{3}$

b $23.8°, 203.8°$

17 a Example: $A = 60°$, $B = 0°$; $\sec(A + B) = 2$,
$\qquad\qquad \sec A + \sec B = 2 + 1 = 3$

b L.H.S. $= \dfrac{\sin \theta}{\cos \theta} + \dfrac{\cos \theta}{\sin \theta} \equiv \dfrac{\sin^2 \theta + \cos^2 \theta}{\sin \theta \cos \theta}$

$\qquad\qquad \equiv \dfrac{1}{\frac{1}{2}\sin 2\theta} \equiv 2\operatorname{cosec} 2\theta =$ R.H.S.

18 c $0, \dfrac{\pi}{18}, \dfrac{5\pi}{18}, \dfrac{13\pi}{18}, \dfrac{17\pi}{18}, \pi$

19 a Expand both sides and use $\sin 30° = \frac{1}{2}$,
$\qquad \cos 30° = \dfrac{\sqrt{3}}{2}$.

  b i Use $\cos 2\theta = 1 - 2\sin^2 \theta$ and
$\qquad\quad \sin 2\theta = 2\sin \theta \cos \theta$.
  iii $26.6°, 206.6°$

20 a $2.5\sin(2x + 0.927)$
  b $\frac{3}{2}\sin 2x + 2\cos 2x + 2$
  c 4.5

Exercise 8A

1 a $8(1 + 2x)^3$ **b** $20x(3 - 2x^2)^{-6}$
  c $2(3 + 4x)^{-\frac{1}{2}}$ **d** $7(6 + 2x)(6x + x^2)^6$
  e $-2(3 + 2x)^{-2}$ **f** $-\frac{1}{2}(7 - x)^{-\frac{1}{2}}$
  g $128(2 + 8x)^3$ **h** $18(8 - x)^{-7}$

2 -1

3 -54

4 $\frac{1}{10}$

5 $5\frac{1}{3}$

Exercise 8B

1 a $(1 + 3x)^4(1 + 18x)$ **b** $2(1 + 3x^2)^2(1 + 21x^2)$
  c $2x^2(2x + 6)^3(7x + 9)$ **d** $3x(5x - 2)(5x - 1)^{-2}$

2 a 52 **b** 13 **c** $\frac{3}{25}$

3 $(2, 0)$ and $(-\frac{1}{3}, 12\frac{19}{27})$

Exercise 8C

1 a $\dfrac{5}{(x + 1)^2}$ **b** $\dfrac{-4}{(3x - 2)^2}$

c $\dfrac{-5}{(2x + 1)^2}$ **d** $\dfrac{-6x}{(2x - 1)^3}$

e $\dfrac{3(5x + 6)}{(5x + 3)^{\frac{3}{2}}}$

2 $\frac{1}{16}$

3 $\frac{2}{25}$

Exercise 8D

1 a $2e^{2x}$ **b** $-6e^{-6x}$ **c** e^{x+3} **d** $24xe^{3x^2}$
  e $-9e^{3-x}$
  f $e^{2x} + 2xe^{2x}$ $= e^{2x}(1 + 2x)$
  g $2xe^{-x} - (x^2 + 3)e^{-x}$ $= -e^{-x}(x^2 - 2x + 3)$
  h $3e^{x^2} + 2x(3x - 5)e^{x^2}$ $= e^{x^2}(6x^2 - 10x + 3)$
  i $8x^3e^{1+x} + 2x^4e^{1+x}$ $= 2x^3e^{1+x}(4 + x)$
  j $9e^{3x} + (27x - 3)e^{3x}$ $= 3e^{3x}(9x + 2)$
  k $e^{-2x} - 2xe^{-2x}$ $= e^{-2x}(1 - 2x)$
  l $\dfrac{2x^2e^{x^2} - e^{x^2}}{x^2}$ $= \dfrac{e^{x^2}(2x^2 - 1)}{x^2}$
  m $\dfrac{xe^x}{(x + 1)^2}$
  n $-\dfrac{(4x + 5)e^{-2x}}{2(x + 1)^{\frac{3}{2}}}$

2 0

3 8

4 $y = 2ex - \dfrac{e}{2}$

5 $y = \frac{1}{3}e$

6 $(0, 0)$ minimum $(2, \dfrac{4}{e^2})$ maximum

7 $\dfrac{e^{3x}(3x - 1)}{x^2}$; $\dfrac{e^{3x}(9x^2 - 6x + 2)}{x^3}$; $(\frac{1}{3}, 3e)$ minimum

Exercise 8E

1 a $\dfrac{1}{x + 1}$ **b** $\dfrac{1}{x}$ **c** $\dfrac{1}{x}$ **d** $\dfrac{5}{5x - 4}$

e $\dfrac{3}{x}$ **f** $\dfrac{4}{x}$ **g** $\dfrac{5}{(x + 4)}$ **h** $1 + \ln x$

i $\dfrac{x + 1 - x \ln x}{x(x + 1)^2}$ **j** $\dfrac{2x}{x^2 - 5}$

k $\dfrac{3 + x}{x} + \ln x$ **l** $e^x\left(\dfrac{1}{x} + \ln x\right)$

Exercise 8F

1 a $5 \cos 5x$
  b $\cos \frac{1}{2}x$
  c $6 \sin x \cos x = 3 \sin 2x$
  d $2 \cos (2x + 1)$
  e $8 \cos 8x$
  f $4 \cos \frac{2}{3}x$
  g $3 \sin^2 x \cos x$
  h $5 \sin^4 x \cos x$

Exercise 8G

1 a $-2 \sin x$
  b $-5 \cos^4 x \sin x$
  c $-5 \sin \frac{5}{6}x$
  d $-12 \sin (3x + 2)$
  e $-4 \sin 4x$
  f $-6 \cos x \sin x = -3 \sin 2x$
  g $-2 \sin \frac{1}{2}x$
  h $-6 \sin 2x$

Exercise 8H

1 a $3 \sec^2 3x$
  b $12 \tan^2 x \sec^2 x$
  c $\sec^2(x - 1)$
  d $\frac{1}{2}x^2 \sec^2 \frac{1}{2}x + 2x \tan \frac{1}{2}x + \sec^2(x - \frac{1}{2})$

Exercise 8I

1 a $-4 \operatorname{cosec}^2 4x$
  b $5 \sec 5x \tan 5x$
  c $-4 \operatorname{cosec} 4x \cot 4x$
  d $6 \sec^2 3x \tan 3x$
  e $\cot 3x - 3x \operatorname{cosec}^2 3x$
  f $\dfrac{\sec^2 x(2x \tan x - 1)}{x^2}$
  g $-6 \operatorname{cosec}^3 2x \cot 2x$
  h $-4 \cot (2x - 1) \operatorname{cosec}^2 (2x - 1)$

Exercise 8J

1 a $3 \cos 3x$
  b $-4 \sin 4x$
  c $5 \sec^2 5x$
  d $7 \sec 7x \tan 7x$
  e $-2 \operatorname{cosec} 2x \cot 2x$
  f $-3 \operatorname{cosec}^2 3x$
  g $\frac{2}{5} \cos \frac{2}{5}x$
  h $-\frac{3}{7} \sin \frac{3}{7}x$
  i $\frac{2}{5} \sec^2 \frac{2}{5}x$
  j $-\frac{1}{2} \operatorname{cosec} \frac{x}{2} \cot \frac{x}{2}$
  k $-\frac{1}{3} \operatorname{cosec}^2 \frac{1}{3}x$
  l $\frac{3}{2} \sec \frac{3}{2}x \tan \frac{3}{2}x$

2 a $2 \sin x \cos x$
  b $-3 \cos^2 x \sin x$
  c $4 \tan^3 x \sec^2 x$
  d $\frac{1}{2}(\sec x)^{\frac{1}{2}} \tan x$
  e $-\frac{1}{2}(\cot x)^{-\frac{1}{2}} \operatorname{cosec}^2 x$
  f $-2 \operatorname{cosec}^2 x \cot x$
  g $3 \sin^2 x \cos x$
  h $-4 \cos^3 x \sin x$
  i $2 \tan x \sec^2 x$
  j $3 \sec^3 x \tan x$
  k $-3 \cot^2 x \operatorname{cosec}^2 x$
  l $-4 \operatorname{cosec}^4 x \cot x$

3 a $-x \sin x + \cos x$
  b $2x \sec 3x + 3x^2 \sec 3x \tan 3x$
  c $\dfrac{2x \sec^2 2x - \tan 2x}{x^2}$
  d $3 \sin^2 x \cos^2 x - \sin^4 x$
  e $\dfrac{2x \tan x - x^2 \sec^2 x}{\tan^2 x}$
  f $\dfrac{1 + \sin x}{\cos^2 x}$
  g $e^{2x} (2 \cos x - \sin x)$
  h $e^x \sec 3x (1 + 3 \tan 3x)$
  i $\dfrac{3 \cos 3x - \sin 3x}{e^x}$
  j $e^x \sin x(\sin x + 2 \cos x)$
  k $\dfrac{\tan x - x \sec^2 x \ln x}{x \tan^2 x}$
  l $\dfrac{e^{\sin x}(\cos^2 x + \sin x)}{\cos^2 x}$

Mixed exercise 8K

1 a $\dfrac{2}{x}$
  b $3x^2 \cos 3x + 2x \sin 3x$

2 $f'(x) = -\dfrac{x}{2} + \dfrac{1}{x}$

4 a $\dfrac{x \cos x - \sin x}{x^2}$
  b $\dfrac{-2x}{x^2 + 9}$

6 $x < -\sqrt{2},\ x > \sqrt{2}$
7 $x < 4$
8 $0.25,\ 1.57,\ 2.89,\ 4.71$
10 a $f'(x) = 0.5e^{0.5x} - 2x$
  b $f'(6) = -1.96,\ f'(7) = 2.56$

11 $\left(\dfrac{3\pi}{8},\ \dfrac{1}{\sqrt{2}}\,e^{\frac{3\pi}{4}}\right)$ maximum $\left(\dfrac{7\pi}{8},\ -\dfrac{1}{\sqrt{2}}\,e^{\frac{7\pi}{4}}\right)$ minimum

12 $y = 2x + 4$
13 a $x = \frac{1}{3}$ **b** $y = -\frac{1}{2}x + 1\frac{1}{2}$
14 b $y = 2x + 1$

15 a $y + 2y \ln y$ **b** $\dfrac{1}{3e}$

16 a $-(x^3 - 2x)e^{-x} + (3x^2 - 2)e^{-x}$
17 a $1 + x + (1 + 2x) \ln x$

Review exercise 2

1 a

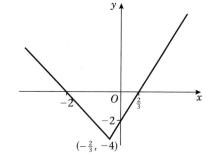

b A: $x = \dfrac{1 + k}{3}$ B: $x = 1 - k$

2 a $y = |3x + 2| - 4$

  b i $x = 1$ **ii** $x = -1\frac{1}{2}$

3 a

M^1(2, 7)

b

M(2, 4)

c

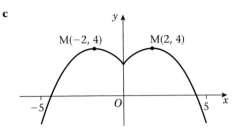

M(−2, 4)　　　M(2, 4)

4 a

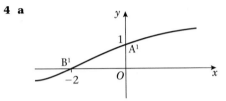

b

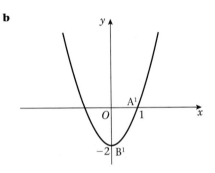

c

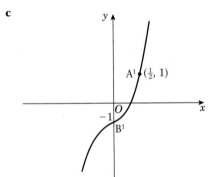

d

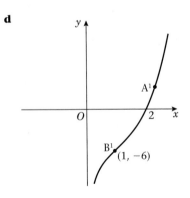

5 a

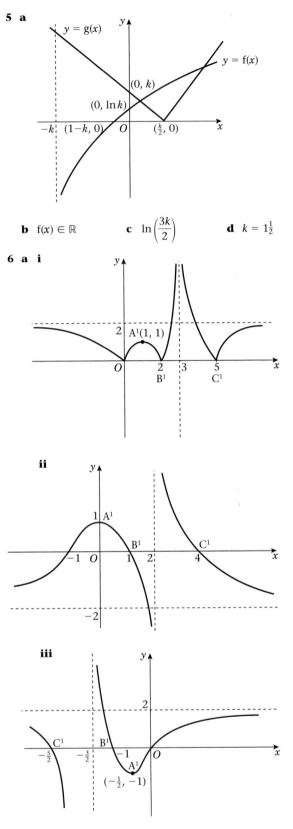

b $f(x) \in \mathbb{R}$　　　**c** $\ln\left(\dfrac{3k}{2}\right)$　　　**d** $k = 1\frac{1}{2}$

6 a i

ii

iii

b i number of solutions is 6
　ii number of solutions is 4

7 a A: (1, 0)   B: (5, 8)

b

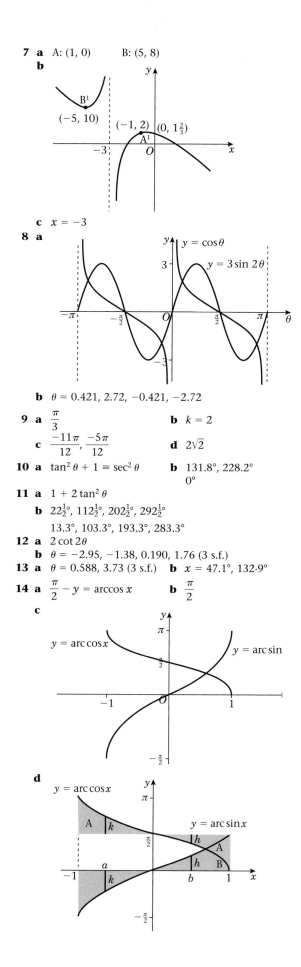

c $x = -3$

8 a

b $\theta = 0.421, 2.72, -0.421, -2.72$

9 a $\dfrac{\pi}{3}$ **b** $k = 2$

  c $\dfrac{-11\pi}{12}, \dfrac{-5\pi}{12}$ **d** $2\sqrt{2}$

10 a $\tan^2\theta + 1 \equiv \sec^2\theta$ **b** $131.8°, 228.2°$
                                                        $0°$

11 a $1 + 2\tan^2\theta$

  b $22\frac{1}{2}°, 112\frac{1}{2}°, 202\frac{1}{2}°, 292\frac{1}{2}°$
    $13.3°, 103.3°, 193.3°, 283.3°$

12 a $2\cot 2\theta$
  b $\theta = -2.95, -1.38, 0.190, 1.76$ (3 s.f.)
13 a $\theta = 0.588, 3.73$ (3 s.f.)   **b** $x = 47.1°, 132.9°$

14 a $\dfrac{\pi}{2} - y = \arccos x$ **b** $\dfrac{\pi}{2}$

  c

  d

The shaded areas A and B are congruent due to the symmetries of the graphs.
Consider $x = a$, where $-1 \le a < 0$

$\arccos x = \dfrac{\pi}{2} + h$, see diagram $\arcsin x = -h$

$\Rightarrow \arcsin x + \arccos x = \dfrac{\pi}{2}$

Consider $x = b$, where $0 \le b \le 1$

$\arccos x = \dfrac{\pi}{2} - k$

$\arcsin x = k$

$\Rightarrow \arcsin x + \arccos x = \dfrac{\pi}{2}$

15 a $4\cos^3\theta - 3\cos\theta$

  b $\sec 3\theta = \dfrac{-27}{19\sqrt{2}} = \dfrac{-27\sqrt{2}}{38}$

16 a $8 + 5\sqrt{3}$ **b** $8 - 5\sqrt{3}$

17 a $\dfrac{-3\sqrt{7}}{8}$

  b i $\cos\left(2x + \dfrac{\pi}{3}\right) + \cos\left(2x - \dfrac{\pi}{3}\right) = \cos 2x$

   ii $\sin 2x$

18 a $-180°, 0°, 30°, 150°$
  b $-58.3°, 31.7°$ (1 d.p.)

19 a $2\operatorname{cosec} 2\theta$
  b

  c $\theta = 20.9°, 69.1°, 200.9°, 249.1°$ (1 d.p.)

20 a $3\sin x + 2\cos x = \sqrt{13}\sin(x + 0.588\ldots)$
  b 169
  c $\Rightarrow x = 2.273, 5.976$ (3 d.p.)

21 $y = -3x + 9$

22 a i $3\sin 2x + 2\sec 2x \tan 2x$

   ii $3\{x + \ln(2x)\}^2 \left\{1 + \dfrac{1}{x}\right\}$

  b $\dfrac{-8}{(x-1)^3}$

23 a $\dfrac{\mathrm{d}y}{\mathrm{d}x} = \dfrac{1}{4}$ **b** $x = \ln 6$

24 a i $(3x + 2)x\mathrm{e}^{3x+2}$

   ii $\dfrac{-[6x^3\sin(2x^3) + \cos(2x^3)]}{3x^2}$

  b $\pm\dfrac{1}{2\sqrt{16 - x^2}}$

25 a $\dfrac{y}{2}\mathrm{e}^{\sqrt{y}}(y + 4)$ **b** $\dfrac{\mathrm{e}^{-2}}{12}$

26 a $\dfrac{\sqrt{3}}{2}$ **b** $\dfrac{1}{\sqrt{1 + x^2}}$

27 a $e^{-x}[-x^2 + 2x]$

   b $e^{-x}[x^2 - 4x + 2]$

   c $x = 0, y = 0$

      $x = 2, y = 4e^{-2}$

   d $(0, 0)$ is a minimum point

      $(2, 4e^{-2})$ is a maximum point

28 a $R \sin 2x \cos k\pi + R \cos 2x \sin k\pi$

      $k = \frac{1}{3}$

      $R = 2$

   b $\tan\left(2x + \dfrac{\pi}{3}\right) = \dfrac{1}{\sqrt{2}}$

29 a $\dfrac{\sqrt{2}\,\pi^2}{18}$ **b** 0.809 (3 s.f.)

   c $x \tan x = 4$ **d** 1.2646 (4 d.p.)

30 a $4x + \sin 4x - 2 = 0$

   b $x_1 = 0.2670$ (4 d.p.)

      $x_2 = 0.2809$ (4 d.p.)

      $x_3 = 0.2746$ (4 d.p.)

      $x_4 = 0.2774$ (4 d.p.)

   c k is 0.277 (3 s.f.)

Practice paper

1 $(e^{-\frac{1}{2}}, -\frac{1}{2}e^{-1})$

2 a 0.092 **b** $\frac{1}{2}(1 + \ln 2)$

3 b $1 + \dfrac{4}{3 - x}$ **c** $x < 3$

4 a

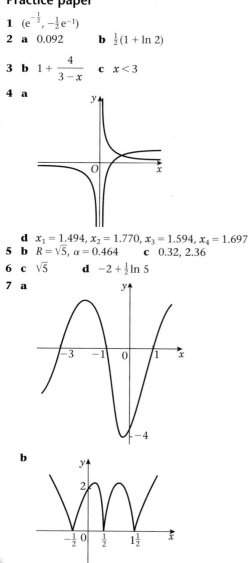

   d $x_1 = 1.494, x_2 = 1.770, x_3 = 1.594, x_4 = 1.697$

5 b $R = \sqrt{5}, \alpha = 0.464$ **c** $0.32, 2.36$

6 c $\sqrt{5}$ **d** $-2 + \frac{1}{2}\ln 5$

7 a

b

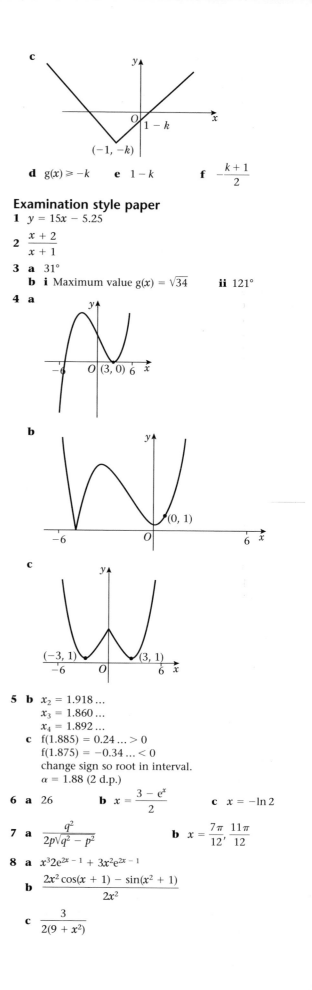

c

   d $g(x) \geqslant -k$ **e** $1 - k$ **f** $-\dfrac{k + 1}{2}$

Examination style paper

1 $y = 15x - 5.25$

2 $\dfrac{x + 2}{x + 1}$

3 a $31°$

   b i Maximum value $g(x) = \sqrt{34}$ **ii** $121°$

4 a

b

c

5 b $x_2 = 1.918\ldots$

      $x_3 = 1.860\ldots$

      $x_4 = 1.892\ldots$

   c $f(1.885) = 0.24\ldots > 0$

      $f(1.875) = -0.34\ldots < 0$

      change sign so root in interval.

      $\alpha = 1.88$ (2 d.p.)

6 a 26 **b** $x = \dfrac{3 - e^x}{2}$ **c** $x = -\ln 2$

7 a $\dfrac{q^2}{2p\sqrt{q^2 - p^2}}$ **b** $x = \dfrac{7\pi}{12}, \dfrac{11\pi}{12}$

8 a $x^3 2e^{2x - 1} + 3x^2 e^{2x - 1}$

   b $\dfrac{2x^2 \cos(x + 1) - \sin(x^2 + 1)}{2x^2}$

   c $\dfrac{3}{2(9 + x^2)}$

Index